THEORY AND MEANING

THEORY AND MEANING

BY

DAVID PAPINEAU

CLARENDON PRESS · OXFORD
1979

Oxford University Press, Walton Street, Oxford OX2 6DP

OXFORD LONDON GLASGOW
NEW YORK TORONTO MELBOURNE WELLINGTON
KUALA LUMPUR SINGAPORE JAKARTA HONG KONG TOKYO
DELHI BOMBAY CALCUTTA MADRAS KARACHI
NAIROBI DAR ES SALAAM CAPE TOWN

ISBN 0 19 824585 8

© *David Papineau 1979*

Published in the United States
by Oxford University Press
New York

British Library Cataloguing in Publication Data

Papineau, David
  Theory and meaning.
  1. Science—Philosophy
  I. Title
  501   Q175   79-40331

  ISBN 0-19-824585-8

*Phototypeset in V.I.P. Garamond by
Western Printing Services Ltd, Bristol
Printed in Great Britain by
The Pitman Press*

PREFACE

This book has developed out of the doctoral dissertation 'A Theory of Meaning for Scientific Terms' which I completed at Cambridge in 1974. The differences between book and dissertation increase as the book goes on. By and large the first two and a half chapters follow the dissertation, matters of presentation aside; while the rest of the book contains both new topics and new conclusions.

Many friends and students have contributed to the development of my ideas about meaning and theory change. I am grateful to them for their interest. Particular thanks are due to my Ph.D. supervisors—Ian Hacking and Mary Hesse—and also to Michael Devitt, Karen Green, Chris Hookway, San MacColl, Ian McFetridge, Hugh Mellor, Philip Pettit, Mary Tiles, and Greg Walkerden, Of course none of these people should be saddled with my views.

DCP

CONTENTS

INTRODUCTION

How sentences in scientific theories are to be evaluated as representations of reality depends on the way that scientific terms acquire meanings. In this book I shall be concerned with those aspects of the theory of meaning for scientific terms which are relevant to questions about the evaluation of scientific theories.

The contemporary debate about theory-choice in science is normally presented as a conflict between two sets of ideas. On the one hand are notions of objectivity, realism, rationality, and progress in science. In opposition is the view that meanings depend on theory, with associated claims about the theory-dependence of observation, the theoretical context account of meaning, incommensurability, and so on. I intend to show that there is no real contest here—that the two sets of ideas are in fact quite compatible. More specifically, I shall argue that the meanings of all scientific terms, including those used to report observations, are inseparable from the total context of surrounding theory and so will inevitably vary with theoretical change, but that this is quite consistent with a broadly objectivist account of science.

The order of exposition will be vaguely chronological. The issues to be discussed will be introduced in roughly the order in which they have emerged in the recent history of the subject. Thus in the first half of the book I shall show how ideas about the theory-dependence of observation and meaning have led to the breakdown of the traditional empiricist account of science, and how some of the more obvious responses to these ideas are inadequate. Then in the latter half I shall show how these ideas can satisfactorily be accommodated within a non-relativist account of science.

One merit of this quasi-chronological approach will be to make it clear that the arguments for theory-dependence are of serious importance for the analysis of scientific discourse, and cannot be dismissed, as is sometimes suggested, as of merely peripheral interest. On the other hand, the approach adopted does mean that a number of points are introduced some time after they first become relevant to the discussion. Thus possible alternatives to a verificationist approach to meaning are not considered until the end of Chapter 3. Again, the distinction between epistemological objectivity and what I call 'semantico-ontological' objectivity, as different targets for the relativist threat, is largely ignored until Chapter 5. And throughout most of the book meaning is treated primarily as an attribute of unstructured terms: only in the last chapter do I come to consider the way the meanings of sentences derive from the meanings of their structural components.

In detail the plan of the book is as follows. In Chapter 1 I start with the 'problem of theoretical terms' and the relativist difficulties that this problem poses for the double language model of science. I then look in detail at the arguments which establish the theory-dependence of observation and undermine the distinction between observational and theoretical language. The upshot of these arguments is that the earlier relativist difficulties about 'theoretical' terms become general. In Chapter 2 I describe how Kuhn and Feyerabend make these difficulties explicit by proposing a 'theoretical context' account of meaning. This proposal is clarified and criticized, but no satisfactory account of the meanings of scientific terms seems forthcoming. Chapter 3 begins by examining Scheffler's suggestion that the problems associated with the theory-dependence of meaning can be avoided by switching attention from the senses to the references of scientific terms. This switch is shown to be of no immediate help. I then consider arguments for equating the meaning of a sentence with the belief it expresses: this involves a lengthy analysis of the problem of interpreting a radically alien language, and of the possibility of conceptual variation across cultures. I show that there are good reasons for adopting a belief theory of meaning, but that these still leave the theory-dependence of meaning and the attendant relativist difficulties untouched. I then argue that meaning involves a specification of truth conditions rather than verification conditions. But this too proves of no assistance with theory-dependence and relativism. In Chapter 4 I turn away from meanings and consider Lakatos's 'methodology of scientific research programmes'. After making various criticisms and corrections, I argue that the resulting methodology is, and can be justified as, a satisfactory account of how to choose between scientific views. But I admit that this account is not so much independent of meanings as implicitly committed to a holist view thereof. The distinction between epistemological and semantico-ontological objectivity is reintroduced at the beginning of Chapter 5, and the priority of the latter explained. I argue for a 'holist' version of realism about scientific theories, rather than one involving one possible truth of constituent sentences, and this holist realism is then shown adequate to account for the possibility of conflicts between scientific views. The causal theory of reference and the notion of partial reference are considered as possible ways of rehabilitating traditional realism, but both prove inadequate. In Chapter 6 I consider the objection that a holist account of meaning for scientific terms cannot accommodate the fact that we understand and decide on new sentences on the basis of our understanding of their constituent parts. The answer to this objection involves a lengthy discussion of 'semantic accounts' of logical form and related topics. I conclude by criticizing Quine's thesis of the indeterminacy of translation and by distinguishing this thesis from the approach to translation which emerges from my holism.

Two omissions need mentioning. Firstly, I assume throughout that all universal generalizations in science are deterministic, and simply ignore the possibility of probabilistic generalizations. This assumption is clearly unrealistic. However I doubt if it matters for any of my arguments. There are of course problems about the precise form and content of probabilistic generalizations. And difficulties about the logic of statistical tests make it unclear exactly what constitutes an 'anomaly' for a probabilistic generalization or a system of such generalizations. But there is no reason to doubt that once these problems are solved probabilistic generalizations will fall quite straightforwardly under the subsequent analysis. Secondly, I nowhere discuss exactly what counts as 'science'. This is because I do not recognize any serious distinction between the natural sciences and other systems of factual thought. In particular I think that the standards of the natural sciences are applicable both to the social sciences (cf. Papineau [1978]) and to everyday thought. However, I do not argue for these claims here, and so will have no objection to those who wish to construe my arguments about 'science' more narrowly than I would myself.

1

THEORY AND OBSERVATION

1 EMPIRICISM AND LOGICAL POSITIVISM

Empiricism has it that all ideas are replicas of sense impressions. In the linguistic mode this becomes the logical positivist principle that any meaningful descriptive expression must be associated with certain sensory experiences.

The empiricist tradition has the virtue of promising straight-forward answers to two central questions. On the *semantico-ontological* question of what our expressions *refer* to empiricism says simply that each expression refers to that entity which is the object of the associated sense impression. And it deals just as comfortably with the *epistemological* question of how we can *know* whether a given statement is true—any statement can in principle be assessed for truth by reference to the sensory experiences we actually have.

2 THE PROBLEM OF THEORETICAL TERMS

The 'theoretical' discourse of modern science poses a prima-facie problem for empiricism. Especially in the last 150 years scientists have made increasingly free with talk of 'atoms', 'force fields', 'mag-netization', 'genes', 'recessiveness', etc. These 'theoretical' expres-sions apparently refer to specifically *unobservable* things and proper-ties. This makes it unclear what relationship such expressions can have to elements of sensory experience. For empiricism, however, it is only association with sensory experience that can give meaning to a descriptive term. So, unless much of the talk of respectable scientists is simply to be declared nonsensical, the empiricist needs to give some further account of how expressions apparently referring to unobserv-ables get meanings.

In this section and the next I shall discuss some of the various ways in which modern empiricists have tried to deal with the 'problem of theoretical terms'. Later on in this chapter we will find reason to be suspicious of the assumptions in terms of which this problem is posed. But our discussion of the problem will remain of value. For one corollary of doubting whether there is a specific problem of theoretical terms is that the difficulties about 'theoretical' terms then arise for scientific language in general.

The simplest and earliest response to the problem was to suppose that theoretical expressions are always explicitly definable in terms of *observational* expressions which do get their meanings from direct association with sense impressions. (The classic expression of this position is Carnap's [1928].) On this view any theoretical expression can always be replaced by some—possibly very complicated—observational expression which is precisely equivalent in meaning. Thus 'force fields' might be defined in terms of the observable accelerations of relevantly placed and constituted material bodies; talk about the 'atoms' and 'electrons' in certain substances could be defined by reference to the observable proportions in which the relevant substances combine chemically; etc. On this view theoretical talk is of no substantial significance: it is merely a useful shorthand for making complex observational statements.

Unfortunately this simple view of theoretical terms does not stand up to examination. An initial difficulty (first made clear in Carnap's [1936]) is that in general theoretical terms are *under*defined in observational language. Take a relatively simple theoretical term like *temperature*. Presumably this term is to be defined by reference to the observable readings on instruments like thermometers. But such a definition will only specify an observational import for the 'temperature' of a body for those special circumstances where a thermometer actually is applied to the body in question. Yet we certainly take it that it is quite meaningful to ascribe 'temperatures' to bodies which are not in contact with thermometers. So our definition is deficient in so far as it does nothing to explain in observable terms what it is for a body to have a 'temperature' in those cases where thermometers are absent.

This apparent underdefinition of theoretical terms in observation language arises quite generally. For most, if not all, theoretical terms are *dispositional* with respect to observations: they specify that certain observable features would be displayed if certain circumstances were to obtain. That an entity does do something observable in actual circumstances is of course a circumstance describable in observational terms. But that it would do something observable in different circumstances, is not itself an observable state of affairs, nor is it clear how it can be reduced to one.

This kind of underdefinability will be relevant to many of the issues to be discussed later in this book. But even more significant will be a contrary difficulty about the observational definability of theoretical terms. For theoretical terms are not only characteristically underdefined in observational language: they are also—in a different dimension—all *over*defined. Consider 'temperature' again. There are in fact a number of alternative possibilities for (under)defining 'temperature' observationally. For we take it that temperatures can be measured either by mercury thermometers or by gas thermometers or by alcohol thermometers or, for that matter, by any number of different devices. Which instruments' readings are then supposed to define 'temperature'? The difficulty is that there does not seem to be any basis for selecting out one of these possible measuring procedures in preference to the others as giving *the* meaning of 'temperature'. (See Hempel [1966], Ch. 7 for a discussion of this and related points; also his [1965], especially Chs. 4–8.)

Why not consider the observational meaning of 'temperature' to be given simultaneously by all the different procedures for measuring it? This makes a kind of sense, but strictly speaking such a 'multiple definition' fails to give an observational equivalent for our theoretical term. The difficulty is that such a 'definition' does more than any definition is entitled to. Accepting it commits us to the general postulate that the different measuring procedures involved will always give the same results when applied to the same body. But a definition cannot have such a consequence—different physical instruments cannot be made to give the same readings by some linguistic fiat.

The point at issue can be brought out more clearly by considering what would happen if we came across a case where two of the procedures built into the attempted multiple definition of 'temperature' actually turned out to give different results when applied to the same body. (This kind of thing does happen. In general scientists at any one time do take it that there are a number of alternative observational ways of deciding the applicability of their theoretical terms. And then they often discover subsequently that what they took to be equivalent procedures give conflicting results in certain cases.) What, in our specific case, is to be said about the 'temperature' of the body in question? It cannot be a matter of simply finding out which kind of instrument is unreliable. For remember that we are supposed to get our grasp of 'temperature' entirely from its observational definition. So there is no further court of appeal when that definition breaks down. In such a case the definition cannot decide what we should say about the body's 'temperature'. Yet note that we should be in no doubt about the appropriate observational characterization of the situation: there is no problem about saying that one instrument gives one reading and the other instrument gives another. This makes it clear that our multiple definition fails to give a precise observational equivalent for 'temperature'—even when we are in no doubt about what has been observed we can still be left at a theoretical loss.

Some philosophers of science accept that theoretical terms in actual scientific languages lack precise observational equivalents, but would argue nevertheless that this is a flaw in scientific languages. This is in effect the position adopted by *operationalists*. Operationalists admit that in practice our theoretical terms are overdefined by illegitimate multiple definitions, but view this as a dangerous deficiency in need of remedy. Observational expressions have clear and agreed meanings. Unless we can have a similarly agreed understanding about what theoretical terms signify there will be a danger that theoretical debates in science will degenerate into confused and irresoluble semantic misunderstandings. If our definitions leave room for disagreement about the 'temperature' of a given body, then what possibility is there of resolving the question? So operationalists recommend that we should reform theoretical language by precise 'operational definitions' which lay down one particular measuring procedure or 'operation' to fix the meaning of each theoretical term. (Operationalism offers no remedy for the *under*definition of theoretical terms. Still, underdefinition does not appear to threaten scientific practice in the same way as overdefinition does. When an overdefinition of a term 'breaks down' there will inevitably be a problem about how to modify the previously accepted meaning. The issue does not arise in the same way for underdefinition.)

However, even if we accept that operationalism offers a real solution to a real problem, there are reasons for thinking the cure worse than the disease.

Consider what serious adherence to the operationalist programme would actually involve. We would no longer be entitled to our 'ambiguous' term 'temperature'. We would have to select from all the alternatives one privileged procedure for measuring 'temperature'. Say we chose mercury thermometers. We would then need another term for what gas thermometers measure. That is, we would need to distinguish 'gas-thermometer temperature' from '(mercury-thermometer) temperature'. And so on, for all the other different procedures we currently take to measure 'temperature'. The same would of course go for all other theoretical terms. And this means we would no longer be able to say, for instance, 'Alcohol boils at 78.3°C at standard atmospheric pressure.' This would fragment into a multitude of distinct generalizations, one for each combination of the current ways of deciding 'temperature', 'atmospheric pressure', and 'being alcohol'. By way of illustration, suppose that for each of these three terms we currently recognize ten different procedures for deciding their applicability. We would then need a thousand generalizations to say what in our present unreconstructed language can effectively be implied by thirty-one (thirty specifying the relevant procedures for applying 'temperature', 'pressure', and 'alcohol', plus the original statement of the boiling point of alcohol).

This schematic example only hints at the damage operationalism would do. Scientific theories as they presently exist are articulated structures of great expressive power. We might conceive of a theory as a pyramidal framework made of struts. The observational terms are the points at which the framework is moored to the ground. The theoretical terms are the points at which the struts making up the framework join. The struts then represent the generalizations linking the theoretical terms to the observational terms and each other. Some theoretical terms will be linked directly to a number of observational terms. Other more abstract theoretical terms, higher up the structure as it were, will be linked indirectly to observational terms through being linked to several of the former theoretical terms. And so placing a strut between theoretical terms up in the air will create a great number of pathways through the framework indirectly linking up various observational points on the ground, the more so the higher up the framework that strut is. Or, to relax the metaphor, formulating an abstract theoretical generalization will systematically imply a wealth of generalizations connecting up observational terms. Thus a statement of the boiling point of some substance will imply that *all* the different tests for temperature will each give a certain answer for anything satisfying *any* of the tests for being that substance. The same point will apply, to an inordinately greater degree, to such basic theoretical assumptions as the Newtonian F = ma or Planck's equation E = hν. It is precisely because theoretical terms do not conform to operationalist strictures and are each linked to observation in a multiplicity of ways that they provide such a powerful and economical mode of expression. If each theoretical term had to be understood as tied to just one observational term, they would no longer be able to facilitate science in the way they do. In the end what the operationalist is effectively asking is that we abjure the use of theoretical terms altogether, for part of what makes theoretical terms distinctively 'theoretical' in the first place is precisely that they are linked simultaneously to a number of observational terms.

3 THE DOUBLE LANGUAGE MODEL

By the middle of this century most empiricist philosophers of science had come to accept that theoretical terms neither are, nor should be, precisely definable in observational language. The resulting orthodoxy was encapsulated in the 'double language model' of science. (The most thorough presentation is in Nagel [1961], Ch. 5.) On this model there are two semi-autonomous languages in any branch of science. On the one hand there is the *observation* language. This can be analysed in the standard empiricist way, with each non-logical expression being associated with some kind of sense impression. Then there is the *theoretical* language. Terms in the theoretical language do not have any directly associated sense impressions to get their meaning from. Still, they are considered to get a kind of experiential content from the existence of certain 'correspondence rules' or 'mixed generalizations'. These are 'mixed' in that they contain both theoretical and observational terms. In effect they specify observable symptoms or indicators for the presence of various theoretical states of affairs. Thus we would as before have general postulates about the relationship between certain instrument readings and temperatures. But the correspondence rules thus relating the theoretical language to the observational language should not now be thought of as giving observational *equivalents* or definitions for theoretical terms. Adherents of the double language model recognized the difficulties facing such definitions. That is, they allowed that in so far as theoretical terms have a dispositional character there will be no observational specification of their meanings for circumstances where those dispositions are not displayed. And they acknowledged that the essential 'pyramidality' of scientific theories means that each theoretical term has to be linked to a number of observational indicators. (It is worth repeating here that the underdefinition and the overdefinition of theoretical terms takes place in different dimensions. A theoretical term can be, and most presumably are, both underdefined and overdefined. It can have a number of potentially conflicting observational indicators and yet still not have any observational meaning for cases where none of those indicators are available.)

Of course the double language model still presents the meanings of theoretical terms as *derived* from the connections between theoretical language and observation language. It is only because of the correspondence rules that we can move inferentially from an observational description of a given situation to a theoretical description thereof. And so it is only by virtue of the correspondence rules that there is any possibility of using inductive or hypothetico-deductive reasoning to evaluate generalizations formulated in theoretical terms. We can think of theoretical terms as getting their meanings by a kind of osmosis—meaning seeps up from the observational ground to the higher reaches of theoretical structure via the correspondence rules. So what meaning a theoretical term has is still dependent on how it is linked to various points on the observational ground. But on the double language model this is no longer a matter of each theoretical term being equivalent in meaning to some given observational expression.

This relatively amorphous account of theoretical meaning offered by the double language model amounts to a retreat from empiricist standards. And by so stepping back the double language model raises certain questions about the seriousness of theoretical talk. In particular it forces us to re-ask in connection with theoretical discourse those two questions to which the more pristine form of empiricism offered easy answers. How can we *know* whether theoretical statements are true? And what are those statements supposed to be *about?*

I shall take these two questions in order. At first sight there might seem no particular difficulty about assessing theoretical statements for truth. As explained above the correspondence rules for a theoretical statement will inferentially connect it with certain observational statements. And so why should we not be able to assess theoretical statements unproblematically, if indirectly, by drawing the appropriate inferences from the observation statements to which they are logically related (these observation statements themselves of course being assessed in the old way by reference to sensory experiences actually had)?

However, when we examine the matter more closely we see that both the underdefinition and overdefinition of theoretical terms raise problems for this procedure.

In the first place, and most obviously, the underdefinition of dispositional theoretical terms means that there is no apparent method of deciding the correctness of applications of those terms to cases where none of the test conditions specified by the relevant correspondence rules obtain. How are we supposed to tell whether something has a certain 'temperature' when there are no thermometers in contact with it and the circumstances are such that no other observable symptoms of the 'temperature' are displayed?

And secondly, and rather more importantly, the overdefinition of theoretical terms also raises epistemological difficulties about theoretical statements. If theoretical terms each have numbers of correspondence rules, then as we saw there is the possibility of observational discoveries showing us that certain of these different correspondence rules are incompatible. When such discoveries are made it will clearly be necessary to revise some of the correspondence rules in question. What principles are supposed to govern the revision of correspondence rules? Which should we stick with when a conflict arises?

One plausible answer is that we should look to our other assumptions involving the theoretical term in question to decide the matter. That is, we can appeal to our ideas relating that theoretical term to other theoretical terms to show which of the contending correspondence rules should be retained. Thus, to borrow an example from Hempel ([1966], pp. 95–6), the earliest ways of measuring *time* were based on the observable periodicities which derive from the daily revolution of the earth. Later, other methods of measuring temporal intervals were added, based on other astronomical regularities and on various mechanical periodicities. But with more thorough and careful experimentation it was discovered that the original measurement procedures conflicted with the later methods. When the diurnal periodicity indicated that two successive time intervals were of equal length, the other methods implied that the first interval was shorter than the second. In practice the conflict was resolved uncomplicatedly enough, by appeal to our basic physical assumptions about the way material bodies change position over time in response to the forces acting on them. For, given these physical assumptions, it makes sense to stick with the later ways of measuring 'time' and admit that the daily rotation of the earth is 'slowing down', rather than to stick with the original method and have to conclude that all the other periodicities in question are 'speeding up': our physics yields an easy enough explanation of the slowing down of the earth's rotation, in terms of tidal friction, but offers no comparable alternative to account for everything else 'speeding up'.

This suggests that in general it will be possible to decide between conflicting correspondence rules for some theoretical term by considering which of those rules fit best with our postulates relating that theoretical term to others. The trouble now is that it becomes quite obscure how the worth of those theoretical postulates themselves is to be assessed. The original thought was that statements in the theoretical language could be evidentially assessed indirectly by examining the observational consequences that flow from them via the correspondence rules. But now it apparently turns out that the selection of correspondence rules is itself to be made by reference to which postulates involving theoretical terms are accepted, and that any putative correspondence rule which leads to observational consequences that do not tally with our theoretical assumptions is for that reason to be discarded. It seems that the evidential assessment of theoretical assumptions is no assessment at all.

We can bring the problem at issue into clearest focus by considering situations in which different groups of scientists actually have conflicting views about which abstract postulates involving certain theoretical terms are to be accepted. Thus, for instance, in the nineteenth century chemists disagreed on Avogadro's hypothesis, on whether the same volumes of different gases contained the same number of molecules. For all that has been said so far there seems nothing to stop the scientists in each camp in such a case each sticking to their favoured position, preferentially selecting such correspondence rules for their theoretical terms (e.g. 'molecule') as are required to substantiate their respective theoretical stances. And the availability of such ploys makes it very difficult to see how anybody could seriously claim to *know* that a postulate like Avogadro's hypothesis is true.

Corresponding to the epistemological difficulties facing the double language model there are semantico-ontological problems about what, if anything, theoretical expressions refer to. I shall now turn to these.

First, again, there is a problem raised by the underdefinition of theoretical terms. If the states of affairs purportedly designated by theoretical expressions like 'being at n°C', or 'being soluble', are ones which it is on occasion in principle impossible to decide the presence of, then what warrant do we have in the first place for construing such terms as referring to possibly undetectable states of affairs? There is a clear enough sense in which things put in water can be 'soluble' or 'insoluble'—but what entitles us to think of things not in water as possessing one or the other of these properties? Or to put the point another way, if we do think of solubility as a property which anything either possesses or lacks, whether or not it is in water, then what warrant is there for supposing that our term 'solubility' as defined by the relevant correspondence rules actually refers to such a property? What indication is there in the way we use 'solubility' for supposing it refers to some enduring state which underlies observable dissolvings?

This question will be raised again in Chapter 3 and in Chapter 6. (It is a question much discussed in the writings of Michael Dummett, first in his [1958] and subsequently by his [1973] and [1976].) Of more immediate import, however, is the way the overdefinition of theoretical terms provides a rather different reason for doubting the ontological status of theoretical discourse. I argued above that scientists appeared always able to retain favoured theoretical statements in the face of awkward evidence by appropriately revising correspondence rules. This arbitrariness about the choice of theoretical statements calls in question whether any such statements can be understood as making claims about determinate aspects of reality. If there is nothing to make one side right and the other wrong when scientists disagree about the acceptability of some theoretical postulate, variously tailoring their correspondence rules to fit, then surely they cannot be held to be disagreeing about objective features of reality? For if they were, if they were talking determinately (if divergently) about certain unobservable entities and properties, then would not the actual relationships amongst those entities and properties make it the case that one side was right and the other wrong?

It is worth being clear about this argument. I am not just repeating the point that it is difficult to *know* whether a certain theoretical postulate is to be accepted, that scientists might always be mistaken in their theoretical assumptions. In general we can allow that decisions on a given kind of statement are always fallible without concluding that there is no reality to which those statements are answerable. Standard sources of error are less than conclusive evidence, or slips in complicated chains of reasoning, etc. We can admit that such sources of error are in practice never eliminable, yet still have a conception of what would have to obtain for a given statement to be true. It might not be within our power to find out whether or not this does obtain, but at least we know what we are looking for. But with theoretical statements we seem to lack even this. Our analysis of the working of theoretical language seems to have left us with no conception at all of theoretical statements being true or false. Scientists seem able to take any theoretical postulates they like and defend them as consistent with the evidence. And it is not just that we cannot show who is wrong; the trouble is that we have failed to attach any content to the idea of anybody being wrong. Which, again, should make us doubt that theoretical talk reports determinately on specific entities and properties—if it did then we could take the correctness of theoretical statements to depend in principle at least on the facts constituted by those entities and properties.

For the most part the adherents of the double language model ignored the related epistemological and ontological difficulties which I have argued arise from the overdefinition of theoretical terms. (Though they did pay some attention to the rather less central difficulties produced by the underdefinition of theoretical terms.) The dominant vision in mid-twentieth century philosophy of science was of science proceeding by rational consensus—it was taken for granted that the canons of scientific rationality would ensure that at any one time scientists would agree as to which views were to be accepted. Active competition between conflicting theories had no particular place in this picture. And so the admitted necessity of adjusting correspondence rules for theoretical terms over time and in accord with new theoretical commitments was not considered to pose any serious threat to scientific objectivity. If new theoretical commitments were only entered into as a result of there being good observational backing in terms of the correspondence rules *then* accepted, then any consequent changes in correspondence rules would be carried out in common by all rational scientists. As intimated earlier, it is precisely when there is a possibility of actively competing theoretical positions that the problems created by the revisability of theoretical terms come into clearest focus. For it is only then that we have the apparent danger that different theoretical camps will mould the observational import of their terms in the different ways required to legitimate their respective theoretical claims. (Both Hempel [1966], pp. 95–7, and Gillies [1972] discuss the possibility of correspondence rules being changed to fit theoretical postulates without apparently seeing any threat to objectivity.)

In recent years the possibility of serious conflicts between alternative theoretical positions has shifted to the centre of our picture of science. I shall discuss the rationale for this shift in the next chapter. For the moment I mention the difference between current and earlier views on this question only as a suggested explanation of the relative blindness of mid-twentieth-century philosophers of science to the difficulties inherent in the double language model.

Not that the adherents of the double language model did not have their own reasons for wondering about the seriousness of theoretical discourse. These arose from doubts about whether there really was any unobservable reality for scientific theories to be about in the first place. Such doubts demand an *instrumentalist* view of scientific theories. According to instrumentalism, scientific theories should not be construed realistically as true or false depending on whether they capture the actual facts about an unobservable reality, but simply as 'instruments' for ordering and summarizing generalizations about observable phenomena. Strictly speaking, theories are not to be analysed as saying anything at all, but just as more or less convenient tools for organizing observational claims.

Note that a refusal to accept the existence of an unobservable reality provides rather different grounds for devaluing theoretical discourse from those I have given above. For I have not questioned that there actually is an unobservable reality. My arguments were rather that, even given that there is such a reality, there are obstacles to our ever finding out about it, and moreover there are reasons for doubting that specific theoretical terms manage to refer determinately to specific aspects of it.

It is worth observing that the adherents of the double language model displayed a strong bias against any realism about theoretical entities. Consider Hempel's much discussed 'theoretician's dilemma':

. . . if the terms and the general principles of a scientific theory serve their purpose, i.e., if they establish definite connections among observable phenomena, then they can be dispensed with since any chain of laws and any interpretative statements establishing such a connection should then be replaceable by a law which directly links observational antecedents to observational consequents. (Hempel [1965], p. 186.)

In response to this apparent paradox various reasons were adduced for not eliminating theoretical discourse. One obvious such reason was indicated by the pyramid metaphor used earlier: theoretical discourse performs the practically essential task of deductively and economically summarizing the complicated mass of purely observational generalizations. But then there is 'Craig's theorem', which shows that, given certain plausible assumptions, for any deductively systematized set of postulates containing both theoretical and observational terms there is another recursively axiomatizable set containing only observational terms but which has the same observational consequences as the original theory (W. Craig [1953], [1956]). The general availability of such purely observational alternative axiomatizations calls in question whether theoretical discourse is really essential even for economical summarization. Not that this argument from Craig's theorem was taken to be conclusive. Some suggested that the alternative axiomatization offered by the theorem would be unmanageably awkward (Hempel [1965], p. 214). Others argued that even if theoretical discourse was not essential for economical systematization, it was nevertheless a prerequisite of scientific growth (Braithwaite [1953]) or of the explanation of observable phenomena (Sellars [1963], Ch. 4) or of certain forms of inductive reasoning (Hempel [1965], pp. 214-15). (See Tuomela [1973] for a thorough survey of the various responses to Craig's theorem and the related results of Ramsey.)

The very terms of the debate about the 'theoretician's dilemma' betray the double language model's underlying prejudice against theoretical realism. For, as Hilary Putnam points out in his [1965], p. 257:

> . . . one can give a very short answer to the question: Why theoretical terms? Why such terms as 'radio star', 'virus', and 'elementary particle'? *Because* without such terms we could not speak of radio stars, viruses, and elementary particles, for example—and we *wish* to speak of them, to learn more about them, to explain their behaviour and properties better.

That is, why should it be necessary in the first place to justify *theoretical* discourse in terms of its ability to assist in the description of *observable* reality? Why should we take it for granted, as Hempel does, that the sole 'purpose' of a scientific theory is to 'establish definite connections amongst observable phenomena'? If observational language is unquestionably legitimated by our interest in describing observable reality, why should we not attempt to justify theoretical discourse similarly in terms of its necessity for describing *un*observable reality? That the matter was not seen like this shows clearly enough the strong predisposition against realism: the idea of justifying theoretical language purely in terms of what it can do for observable talk only makes sense if one starts off with a fundamental suspicion of unobservable reality.

Still, it is not difficult to understand why the double language model should have fomented this initial distrust of unobservables. In the double language model theoretical discourse is forced into an essentially derivative role. Theoretical expressions are meaningful, and statements made using them are testable, only by virtue of their connection with the observation language. And this makes it very natural to view theoretical language as at best only a secondary adjunct relative to the real business of science: namely, to 'establish definite connections amongst observable phenomena'.

4 OBSERVATIONS AS JUDGEMENTS

It is now time to question the central assumption which prejudices the double language model in favour of instrumentalism, and which indeed is a prerequisite of the choice between instrumentalism and realism being raised at all: namely, the idea that in any area of science there are two separate languages, an 'observational language' and a 'theoretical language'. Throughout the last section I went along with this assumption, and correspondingly assumed that 'observation language' would be free of the ailments afflicting 'theoretical language'. But why should we assume there is any such unproblematic observation language?

The idea was that observational expressions get their meanings from direct association with elements of direct sensory experience: statements made using such expressions are consequently supposed to be authoritative descriptions of observable features of reality. But once we stop to examine this idea (as adherents of the double language model characteristically did not—see Hesse [1974], pp. 9–10) we shall find good reasons for doubting whether there are in fact any such observational expressions, and indeed whether there is any philosophically important distinction to be made between observational and theoretical language at all.

An initial difficulty is that many sensory experiences seem to depend on the observer's having certain general hypotheses about the nature of the thing observed. To take a well-known example of N. R.

Hanson's, what the trained physicist sees when presented with an X-ray tube is quite different from what the uninitiated layman will experience. The layman will see a glass and wire thing, like a funny light bulb, but the physicist will see a distinctive piece of equipment with certain definite uses. (Hanson [1958], Ch. 1.) Again, what the layman watching a cloud chamber will see as indeterminate squiggles, the experienced physicist will recognize straight off as certain interactions involving sub-atomic particles. The phonetician or socio-linguist will be able to make auditory distinctions between phonemes and accents which are indistinguishable to the man in the street. Geologists can visually identify rock formations in a way impossible for the untrained. And so on. What these examples seem to show is that the sensory experiences of an observer will depend on his previous introduction to certain specialized assumptions about the thing observed.

If this is right, then sensory experience cannot be a matter of direct and unproblematic contact with observable reality. If different intellectual histories lead to different sensory experiences, how can those experiences be matters of direct access to a common reality?

But perhaps the examples are less than convincing. Can the fact that something is an *X-ray tube*, or an *electron-positron pair*, really be something given directly in sensory experience? There is room to doubt this. Sensory experience can show us the shape, the colour, the hardness, etc. of things—but surely it does not show us how they are to be analysed in terms of the abstract concepts of modern physical science. Thus it might be argued that the judgements in the above examples are not after all parts of direct observational experience, but on the contrary are arrived at only after we move from the realm of the observable to that of the unobservable, only after we shift from observational discourse to theoretical discourse. And similarly with the other examples: it could plausibly be maintained that the dependence of the claims in question on the observer's prior intellectual histories arises only for theoretically inferred conclusions. As long as we stick to observable matters properly so called, then there remains no reason to deny that the sensory experiences of healthy observers put them in common contact with a common reality.

The trouble with this line of defence is that it is doubtful whether sensory experience ever gives us pure and uncontaminated contact with anything. The implicit motivation for the double language model's separation of observational and theoretical statements was of course the thought that the one thing we can be sure of is the evidence presented to us by our senses. But is sensory experience best conceived of on the model of evidence being presented? A number of considerations argue for the alternative view that sensory experience is in essence a process of *judgement*. The sensory experiences we have are invariably more than our sense organs entitle us to, in a way I shall explain below. And the best way of making sense of this is to think of those experiences themselves as being hypothetical judgements (which can fail) rather than as data (which just are). (On this and a number of subsequent points I am following E. Craig [1976]. Other works containing not dissimilar views on observation are Hanson [1958], Ch. 1; Feyerabend [1965a]; Kuhn [1962], Ch. 10; Gregory [1970], [1974].)

Duck-rabbits, Necker cubes, and similar gestalt-switching figures illustrate the basic point I wish to make. With such figures we get a constant input to our sense organs; but we nevertheless switch back and forth between seeing the drawing as a rabbit and seeing it as a duck, between seeing a cube one way up and seeing a cube the other way up. Somehow it seems that what we are immediately aware of in these sensory experiences has a content which adds to what we get by way of sensory input from external reality. The experiences seem to involve hypotheses about the thing experienced which go beyond what our sensory stimuli can guarantee.

It might be thought that these cases are oddities. After all, most of our everyday experiences do not switch back and forth in this way. But the same points can be made about cases that do not involve the switching effect. The 'Ames room' is a good illustration (see Gregory [1970], pp. 26–9). To an appropriately-placed observer an Ames room looks like a normal rectangular room. But it has the odd feature that to (almost) all observers someone walking from one far corner of the room to the other apparently doubles in size. The explanation is that the room is not in fact normal. Its floor plan is really ⌐⌐, rather than ⌐⌐, and in elevation its back wall is really ⌐⌐, rather than ⌐⌐, with the other walls, furniture, and fittings arranged to match.

The crucial point for our purposes is that, despite their 'visual' indistinguishability we do not see the room as one which might either be a normal room or an Ames room. If that were what we saw then presumably our sensory awareness would be correspondingly neutral between the walker getting bigger, and his staying the same size and getting closer. But our visual experience is unequivocally of the walker getting bigger. So we unquestionably (albeit mistakenly) *see* the room's two back corners as having the same height, rather than as being unequal in height, or as one or the other.

Richard Gregory has argued from such examples as the above that we should conceive of our sensory systems as making hypotheses (Gregory [1974]). Our visual system, for instance, transforms an essentially 'two-dimensional' input into a conclusion about the presence of three-dimensional physical objects of certain kinds. We might also say that the visual brain adopts the general postulate: *that* kind of stimulus means we are in a normal rectangular room. In the same way the other sensory systems can be thought of as using general principles to extrapolate from peripheral stimulations to judgements about features of the external world.

This account of sensory experience allows a neat explanation of the kind of switching effect we get with Necker cubes and duck-rabbits: the switches occur because the general postulates 'built into' our visual system support to the same degree two different conclusions about the object of our sensory awareness. For the visual system the alternative 'conclusions' that the Necker cube is one way up and that it is the other way up are equally likely—so it opts first for one and then for the other in random alternation.

It is important to be clear about this model. The suggestion is not that we automatically infer certain conclusions *from* the basic data of sensory experience; that sensory experience is always and inevitably accompanied by a further process of 'interpretation' or 'perception'. On the contrary, the idea is that our sensory experiences *are* the conclusions in question. The inferential interpreting involved does not follow our initial conscious awareness, but precedes it. Consider the Ames room again. The judgement that the room is a normal rectangular one is indubitably in our sensory awareness, and not something additionally inferred in thought. For it is surely uncontroversial that so far as his further judgement goes, any observer will know full well that people do not double in size just from walking across rooms (or from any other cause for that matter). So he will be in no doubt that he is being subject to some kind of illusion. If he doesn't believe at any further level that the walker is getting bigger, any judgement to that effect can only be in the visual experience itself.

It might be objected, in connection with the Necker cube, say, that surely the judgement that the cube is the one way up (or the other) is something additional to our most basic sensory awareness. Do we not first see a certain pattern of lines in a two-dimensional plane, and then go on to judge that it is a cube a certain way up? Well, it is true that we can see the cube as a two-dimensional pattern of lines (and this is so whether it is a drawing on paper or a real cube made of wires). But this experience is not something always coexisting with the other ways of seeing it; something common to the two alternative views of it as a cube. It is another alternative to those views, incompatible with them, just as they are incompatible with each other. In general, it is usually possible with some concentration to see any visual scene as a two-dimensional pattern of textures and colours rather than as a three-dimensional arrangement of physical objects. But again, this is not a more basic kind of awareness always present in any visual experience, but an alternative experience incompatible with what we normally see: it involves the judgement that what is before our eyes actually *is* a two-dimensional pattern of colours, like paint marks on a canvas, which of course contradicts the judgements in our normal visual experience.

The view that sensory experiences are essentially hypothetical judgements poses a serious problem for the conception of observational language embodied in the double language model. If what we are 'given' in sensory awareness is not a direct grasp of certain features of the external world, but always a hypothetical conclusion extrapolated from a fragmentary basis, then the authority of such statements as get their meanings from associations with sensory experience cannot be taken for granted. For there is always the possibility that the sensory experiences which give rise to such statements might be mistaken.

I do not of course want to say that all sensory experiences have to be mistaken; that there is no possibility of a sensory experience being a veridical representation of reality. That our sensory awareness consists of judgements which are 'inferred' from the 'data' via certain general 'assumptions' by no means shows that those judgements cannot be sound. In nearly all cases our visual brains are quite right when they 'conclude' that we are in normal rooms. But the fact that the output of our sensory systems goes beyond what their input can guarantee does mean that we cannot ignore the possibility that any sensory judgement might be mistaken. On the alternative picture, which has our sensory experience giving us direct access to certain features of the external world, there is no question about the reliability of our sensory experiences in informing us about those features. But on our present view it will always make sense to ask, of any given sensory experience, whether we should accept what it tells us or not. However it is to be answered, the very posing of this question undermines the idea of an automatically self-guaranteeing observation language.

At this point a defender of the notion of an authoritative observation language could perhaps object that we are supposing too simple a model of how observational expressions are given a meaning. He could point out that the kind of examples which purport to show that sensory experiences are fallible judgements only work because we can tell whether the situation observed corresponds to the sensory judgement or not. We could only know that a visual judgement to the effect that something is a rectangular space is mistaken if we have some effective alternative method of discovering that the room so visually judged is not rectangular after all. And presumably this method will somehow involve some kind of further observations of the space in question (such as feeling it, measuring it, etc.). So perhaps all we should conclude from the above arguments is that the way observational expressions get their meaning is more complicated than we might unreflectively be inclined to assume. There is a sense in which there is a natural association between the meaning of 'rectangular' and certain visual experiences. But what the 'fallibility' of those experiences suggests is that this association does not itself exhaust the meaning of 'rectangular'. For what really warrants describing a space as 'rectangular' is not just the occurrence of a certain visual experience, but the outcome of some more complex observational investigation involving not just sight but also surveying with rulers, etc.

Perhaps this point should have been obvious from the start. Every competent English speaker knows for instance that the visual judgement that a stick is bent does not in itself warrant applying the word 'bent' to it: a stick which looks bent when partly immersed in water can well be straight. So surely it must be part of the meaning of 'bent' that the relevant visual experience warrants its application only if further observable conditions are satisfied. In general, it would seem that observational expressions do not get meanings just from simple associations with possibly illusory sensory judgements, but from more sophisticated observational procedures which provide built-in checks on the occurrence of such illusions.

This elaboration holds some promise of restoring the conception of an authoritative observation language. But in the end it fails to do so: it turns out to be of no real help to replace the obvious and natural associations between certain words and certain elements of sensory experience by more complicated and less obvious ones. For in so far as sensory experiences, however complex, have the nature of inferred judgements, then there is an important sense in which all observational judgements are *theory-dependent*.

5   THE THEORY-DEPENDENCE OF OBSERVATION

The idea that observation is theory-dependent is much discussed but little understood. As a preliminary, some terminological clarification will be useful. In the previous section I followed the double language model in equating 'theoretical' with 'unobservable'. But if, as I am arguing, there is no serious distinction to be made between what is an observable matter and what is not, then this usage becomes pointless. So from now on I shall use 'theory' in a wider sense to refer to any (lawlike) generalization or interrelated set of such generalizations.

The theory-dependence of observation is often taken to be essentially to do with the phenomenon alluded to right at the beginning of the last section—the possibility that people accepting different theories will as a result have qualitatively different sensory experiences. Now some writers have suggested that any authoritative observational language is impossible because all observational experiences are so theory-dependent. This claim seems to be implicit in Hanson [1958], Ch. 1, and in Kuhn [1962], Ch. 10; and it is quite explicit in Paul Feyerabend's [1965a]. At the end of this article Feyerabend is concerned with the problem of how observational experience can provide any critical evaluation of theories, if what we observationally experience depends in turn on what theory we accept. He suggests the following solution:

. . . our acceptance of very general points of view . . . is based upon the identification of two types of behaviour, viz., (1) the 'behaviour' of certain selected sentences that are predicted by the theory, . . . and (2) the behaviour of those very same sentences as uttered by a human observer who *does not know* the theory. . . . [A general point of view] is removed if it produces observation sentences when observers produce the negation of those sentences. (pp. 214–15.)

The idea is simple enough: a theory is to be repudiated if its predictions are not borne out by observation. But the fact that Feyerabend found it necessary to specify that the observations should come from people who do *not know* the theory being evaluated shows that he actually did think that scientists accepting a theory will be incapable of having sense experiences which contradict it.

However, as Feyerabend himself has since come to realize (Feyerabend [1975], p. 133), it is most implausible to suppose that our sensory systems are entirely at the mercy of our theoretical presuppositions. Even if our theories affect our sensory experiences to some extent, it by no means follows that they do so to that extent.

Recall the model suggested earlier, of our sensory systems using 'general postulates' to infer observational conclusions from limited data. Now it does indeed seem likely that the 'general postulates' so used are not all innate to us as biological organisms, but are to some extent learnable and unlearnable. In particular it seems that in some cases our explicit theoretical postulates will get 'internalized' into our sensory systems in such a way that the conclusions they warrant become part of our immediate sensory experience. Thus the physicist whose explicit theory tells him that certain shaped tracks in cloud chambers are electron-positron pairs will with practice learn to see electron-positron pairs as clearly as he sees anything. The explicit postulates which originally allowed him to consciously compute the nature of the particle interaction from measurements of the pattern of the tracks get built into his unconscious visual system and allow him to tell straight off what kind of interaction has occurred.

But even given this model there is no reason to suppose that our sensory systems are completely plastic with respect to what 'general assumptions' get built into them. There are surely some such 'assumptions' which it is impossible to unlearn. Try as we might, we do not seem able to get ourselves to see straight sticks as having the same shape when half in water as when entirely out. And, on the other hand, there are no doubt some assumptions which are impossible to internalize in the first place: it seems unlikely that we could train ourselves to have direct visual experience of, say, whether the number of people in a fairly large crowd is odd or even. There are no doubt good physiological explanations for both these kinds of limitations. But whatever the appropriate explanation there seems little reason to doubt that the 'assumptions' implicitly built into our visual systems can be different from the theoretical postulates we explicitly accept. And so, *contra* the early Feyerabend and others, it is perfectly possible for the judgements which issue from our sensory system to conflict with our explicit theories.

Still, there is a rather different and rather more important sense in which all observation can be considered theory-dependent. Even if the observational experiences we actually have are not entirely dependent on our explicit theories, it might still be that what we *say* in response to our sense experience is never independent of the theories we accept.

To illustrate this point, suppose that we apply some term to something predictively on the basis of our existing theories, but when we actually observe it our sensory experiences are such as to lead us to deny that term of that thing. Traditionally this would be taken to discredit the theories on which the prediction was based. But why should it? On our present understanding, the sensory experiences in question are the results of 'inferences' based on certain 'general assumptions' built into our sensory systems. Why should we trust those 'inferences' more than the inferences derived from our explicit theories? Is there not as much reason to suppose that the 'assumptions' built into our brain are wrong as to suppose that our theory is?

And indeed, when we think about it, it is a not infrequent occurrence for our explicit theories to teach us that certain sensory experiences are not after all the reliable indicators of external states of affairs that we previously took them to be. Thus the Copernican revolution showed us that the earth was not, despite appearances, standing still. Similarly we have learnt from our electromagnetic and cosmological theories that an impression of redness does not, as we hitherto supposed, necessarily mean that the object observed actually is red (as with distant receding stars). Other examples are not hard to find.

We can understand such cases in the following terms. At one time a certain set of sense experiences are associated with a certain word, in the sense that their occurrence is generally taken to warrant the word's application. But then we discover by reference to our theories that those experiences do not warrant the word's application after all, that they can well occur in cases where the word is not applicable. It is in this sense that what we say in response to sense experience can depend on our theories, even if the sensory experiences in question do not do so.

It should be clear that for theory-dependence in this sense it is of no consequence whether words used observationally are analysed in terms of their natural associations with simple elements of sense experience or in terms of more complex associations which provide a check on the occurrence of recognized illusions. For even given the more elaborate model it will still be perfectly possible for our theories to show us that we were mistaken in taking certain observational procedures to validate the application of some term. In a sense what our theories will then show us is that in addition to the recognized illusions related to the use of some term there are some further, hitherto unrecognized, illusions. We always knew that the observational use of 'red' required that we watch out for funny lighting and rose-tinted glasses—but we had to discover that the relative speed of the object observed might also need taking into account.

When we do discover that our current tendency to use some word in response to certain sensory experiences can lead us astray, a number of different things can happen. We might simply stop using the word observationally altogether, deciding that there was no straightforward way of deciding its applicability on the basis of sensory experience. Alternatively we could possibly alter the set of sense experiences which we recognized as evidencing the word's applicability, by eliminating or replacing those sense experiences in that set which were responsible for the mistaken judgements. And there of course remains the possibility that we might leave just the same set of sense experiences associated with the word, and instead 'learn' to have those experiences in appropriate rather than inappropriate circumstances. This would then be a case where our experiences themselves and not just what we said in response to them depended on our theories. It would be a matter of an explicit theory which conflicted with experiential usage getting 'internalized' into the appropriate sensory systems in such a way as to stop the awkward experiences occurring. Hanson suggests, for instance, that Kepler resolved the potential conflict between his theory that the sun was stationary and the dawn experiences which inclined him to say the sun was 'rising' by coming to see sunrises differently—by coming to see them in terms of the horizon falling. (Hanson [1958], Ch. 1.)

Not that it matters much whether Hanson is right about this particular case. There is indeed a real question for him to be wrong about. The retinal stimulation will of course be the same in either case, but the judgemental theory of perception shows how there can still be a real difference between seeing the sun rising and seeing the horizon falling. My own intuitions are that Hanson is wrong to claim that Kepler's actual sensory experiences changed. But, as I said, this is in itself of no great importance. For we can now see how it would in effect have come to the same thing for Kepler simply to have decided to change the way he was going to use 'stationary' so that his original dawn experiences no longer counted against its application; that is, for him to have decided that any experiences of the sun rising above the horizon should simply be recognized as a kind of illusion. Whichever way the conflict was resolved, whether or not it involved actual experiences changing, it still seems that Kepler's use of the term 'stationary' in the relevant context was dictated as much by his theory that the sun does not move as by any established association between the term and certain sense experiences.

This then illustrates again the general point I wish to make. The crucial issue is not whether our sense experiences themselves are unrestrictedly responsive to our explicit theories. That they are so indefinitely malleable is most implausible (as in particular is Hanson's suggested analysis of the Kepler case). But it might still be that what it is right to say in response to sense experience is always dependent on our theories.

It is this that seems to me to constitute the most serious criticism of any empiricist conception of an authoritative observation language. The original hope was that associations with sense experience would guarantee the authority of certain statements. But it now seems that the ability of certain statements to issue from their associations with sense experience does little in itself to guarantee their accuracy. For there seems always to be a question of whether we are right to associate those sentences with those experiences (or, alternatively, whether we are right to have those experiences when we do). The way we associate words with experiences seems always to require assessment by reference to our general theories. I am not denying that certain sentences do as it happens get asserted in response to the occurrence of certain sensory experiences. The point is that this facet of their use does not seem to determine what those sentences mean, and so does not guarantee their accuracy. On the contrary, that the sentences are so used seems itself answerable to our theories involving the expressions in them, and there is always the possibility that those theories will overturn those observational usages. So even if certain sentences are at a given historical juncture characteristically prompted by the occurrence of certain observational sense experiences it does not seem that this makes them *observation* sentences in the sense desired.

6 OBSERVATION CONTINUED

At this point I shall consider briefly two remaining ploys that might be tried by someone who still hoped to distinguish an authoritative observation language. Firstly there is the possibility that the proper place to look for observational expressions is not in talk of how things are but in talk of how they appear. And then there is the strategy of analysing observational language without bringing in any talk of subjective sensory experiences at all. I shall take these in turn.

Throughout this section I have been assuming that if observational expressions are to be found anywhere it will be in the language we use for describing physical things and properties thereof ('red', 'rectangular', etc.). But there is another mode of describing phenomena: there is the language of how things look, how they appear. Perhaps it is expressions like 'looks rectangular', 'looks red', etc. that make up observation languages. Perhaps statements about how things look can in themselves have an authoritative status, even if talk of how they additionally are is not similarly automatically guaranteed.

This hope was of course the basis of the influential (indeed predominant) *phenomenalist* stream in early and mid-twentieth-century empiricism. For phenomenalists the objects of our direct sensory awareness were sense data. To say that something 'looked red' (or more generally to say that it presented a certain sensory appearance) was to claim that there existed objectively a certain sense datum; and since we had direct access to such sense data such statements could be regarded as indubitable. The phenomenalist programme was then to show how normal everyday talk about physical things and their properties could be 'reduced' to complex constructions of statements about the sense data that constituted the basic material of reality.

However, there are a number of serious objections to this phenomenalist conception of an authoritative observation language. In the first place there are difficulties with the proposed reduction of 'physical thing' language to 'looks' language. The problems which arise are exactly analogous to those which were pointed out earlier in this chapter as facing any attempt to relate talk about unobservables to a putative observation language. Any attempt to carry out the phenomenalist reduction faces the problem that terms for physical things seem to be both overdefined and underdefined relative to the 'looks' language. And this means that there will be difficulties about both the epistemological and the ontological status of statements in the physical thing language. Now it is one thing to say that we cannot ever know about 'atoms' and 'viruses', and that we have no reason to suppose that there is anything in reality answering to our use of those terms. But it is far more difficult to stomach the corresponding thesis about 'tables' and 'chairs'. Talk of 'looks' loses much of its attraction as a possible observation language if it means that we shall have to regard all talk of everyday physical objects as entirely suspect.

Then there is the point that the same thing as seen from the same perspective can look different to different people, in particular to people with different theories. This kind of theory-dependence of experience is not, as we have seen, particularly puzzling in itself. But it is in serious tension with the notion that our primary and most reliable access to reality is via our knowledge of how things look. For it then seems to undermine any idea at all of an objectively existing reality—each individual would have an idiosyncratic personal reality constituted by his own distinctive experience of how things look.

And finally, and more generally, the arguments for a judgemental theory of sense experience suggest that there is something fundamentally mistaken about the phenomenalist understanding of 'looks' talk. For the phenomenalist the looks of things, and their appearances in general, are conceived of as independently existing facets of reality. But, given the judgemental account of experience, to talk about how things 'look' is not to describe certain independent existents to which we have visual access, for sensory experience is not a matter of access to data in the first place. Rather, to talk about how something looks is simply to specify the judgement present in our visual experience. To say that something 'looks red' is to note that we have a visual judgement to the effect that it is red; it is not to ascribe any special kind of feature ('a red look') to that thing. We can think of a statement as to how something looks as an acknowledgement that our visual brain inclines us to a certain belief about that thing. Such an acknowledgement is quite compatible with our otherwise withholding judgement on, or even denying, the belief itself. So any special authority that might be accorded to statements about how things look scarcely makes them observation statements. In a sense such statements fail to say anything at all. If they are infallible it is not because they report matters to which we have special access but simply because they limit themselves to evincing a certain kind of inclination to belief without endorsing it. (Cf. Quinton [1955], and also E. Craig [1976], where Quinton's analysis of 'looks' statements as diffident expressions of belief is discussed and improved on.)

I turn now to the possibility of distinguishing an observation language without bringing in conscious sensory experiences at all. Why not simply count as observational expressions those expressions which we are trained to use unthinkingly and immediately in direct response to the reception of certain physiological stimuli at our sense receptors? Non-observational expressions will then be those which we apply only indirectly, after time and thought, as when we use consciously accepted theoretical postulates to move inferentially from direct observational judgements about a particular situation to its unobservable characteristics (cf. Feyerabend [1958]).

One motivation for this line of approach might be a behaviourist distrust of talk of such 'subjective' entities as conscious sensory experiences. There seem to me no good grounds for this mistrust (stemming as it does in large part from what I am now showing to be the mistaken supposition that talk about observables, such as overt behaviour, is scientifically superior to talk about unobservables, like mental states). But we can consider the merit of the current suggestion quite independently of any behaviourist attractions it may have. Our earlier attempts to distinguish an observation language on the basis of associations between certain expressions and conscious sensory experiences all failed. The current suggestion has the advantage that even if there do exist subjective sensory experiences to be associated with the use of certain words, such associations would in themselves be quite irrelevant to the observational status of those expressions. What makes a term observational is simply our being conditioned to use it directly when the external world interacts with us in a certain physiological way—that certain conscious experiences might accompany such uses would be beside the point. And what then gives different observational terms their distinctive meanings would not be any associated conscious experiences but simply the different peripheral physiological stimuli to which their uses are variously conditioned. (Cf. the notion of 'stimulus meaning' in Quine [1960], Ch. 2.)

However this latest attempt to distinguish an observation language must also be deemed a failure. Here the difficulty is again that there is no reason to take the 'observational' statements thus picked out as having any special reliability in reporting on reality. There is nothing wrong with the picture of our being trained on the one hand to use expressions directly ('observationally') in response to physiological stimuli and on the other to use expressions indirectly ('non-observationally') via inferences based on our theoretical postulates. But why should we accord any authority to statements issuing from the former usages over those issuing from the latter? The arguments against some of our earlier attempts to analyse observation language apply even more directly here. On the sense experience account there was at least the initial hope that observation statements could be construed as expressing some kind of direct access to reality. But if what distinguishes observation reports is simply that they are behavioural responses to physiological promptings then there seems no special reason to rely on them at all. Consider again the situation where an observation report conflicts with an assertion which is independently derived from some theoretical chain of reasoning. Surely we have even more reason than before to construe such cases as simply showing that there is something wrong with the observation report in question. If there is nothing more behind the observation report than a tendency to respond to certain stimuli with a certain form of words, then why should we not just conclude that the tendency has here led us astray, and make efforts to modify it accordingly? And if it can be right to do so, as it manifestly often is, then it seems that our tendency to use certain words in response to physiological promptings does not in itself fix what those words mean, and so does not underwrite their accuracy.

So switching from conscious sense experiences to peripheral stimuli gets us no closer to an empiricist observation language. Indeed there are in any case good reasons against making this switch. Recall that an existing tendency to use words in response to sensory promptings seemed to be alterable in *two* ways: we could either associate these words with different sense experiences, or we could come to have those same sense experiences in different circumstances. However, if we are to analyse the observational use of words purely in terms of peripheral stimuli then these two kinds of alteration come out alike as both simply being changes in the stimuli which prompt the use of the word. In so far as the distinction in question is one worth making, we thus have a reason for sticking with the sense experience account of observational usage. On the other hand, nothing I say henceforth will depend on the distinction in question, and so I shall simplify the subsequent exposition by taking the observational use of words to be just a matter of verbal response to peripheral physiological stimuli.

Let me briefly sum up the discussion of observation and observation language in the last three sections. As has previously been remarked, there is clearly no reason to deny that people make 'observation reports', in the sense that they are on various occasions prompted by their senses to make certain statements. But what is in question is whether the expressions used to make such observation reports form an observation language in the sense this is understood in the empiricist tradition. For it seems that such observation reports have no special authority. We can well discover by reference to our theories that certain such observation reports are mistaken, and that it is correspondingly necessary to alter the way we link the expressions involved to sensory inputs.

Perhaps at this point it might be felt worthwhile to specify a new sense of 'observation language', and to count as part of that 'observation language' all expressions used in observation reports, without any suggestion that expressions so used have any special authority. Well, we could do this, but such an 'observation language' would fail to bear even a superficial resemblance to that envisaged by the double language model. For many, if not all, putatively 'theoretical' expressions ('X-ray tube', 'electron-positron pair') are on occasion used by suitably trained observers to make direct observation reports, and so would be part of the 'observation language'. So the suggestion that any term usable in an observation report is to be part of 'observation language' would leave very little indeed that was not.

A more useful usage, and one I shall adopt henceforth, is to count a term as (relatively) 'observational' if it is used to make observation reports *more often* than it is used on the basis of inferences. Thus 'red' could be counted as more observational than 'ionized'. But it is worth repeating, at the risk of labouring the point, that a term's being observational in this sense does not accord any special authority to observation reports made using it. Even terms like 'red' can be used inferentially as well as observationally, as when we judge that something is 'red' not as a result of looking at it but from knowing the wavelength of the light it emits or from knowing it is iron at 1000°C. And, as we have seen, even for terms like 'red' such inferential uses turn out no less authoritative than observational uses in cases of conflict.

What now of the 'pyramid' metaphor of scientific theories? As originally presented, this was premissed on the double language model's distinction between theoretical and observational language: the points at which the structure was moored to the ground were the 'observational' terms, and the joints in the structure were the 'theoretical' terms. But with appropriate modifications this metaphor can still remain useful. Suppose we forget about the distinction between 'observational' and 'theoretical' terms. Then, as I have suggested, we may as well understand a theory as any structure of interrelated generalizations. But it remains an important point that scientific theories have a significant degree of *deductive systematization*. A scientist does not accept all the generalizations involved in his theory one by one in a long list. Rather there will be some basic set of premises (axioms) which imply all the other generalizations his theory commits him to. (This is something of an idealization, but it is harmless enough in the present context.) Amongst the premises in a deductive systematization some will be more *central* than others, in that they will play a part in implying more derived generalizations. Thus, for instance, 'F = ma' would be more central than Hooke's law in any systematization of classical mechanics. 'F = ma' will be amongst the premises required to deduce almost any derived generalization involving motion in classical mechanics—but Hooke's law will be needed only for those derived generalizations concerning springs. My suggestion is that we should retain the pyramid metaphor by now considering the struts at the highest and narrowest part of the pyramid to be, not the least observational, but rather the most central of the premises in our deductive systematization, with the centrality of premises correspondingly decreasing as we move down the pyramid to the broad base.

The pyramid metaphor was always meant to capture the deductively systematic aspect of scientific thinking, with the struts representing an economical set of premises, and the various possible indirect pathways which the structure allows between the points (expressions) at the end of the struts representing the derived generalizations which can be deduced from those premises. But I originally presupposed implicitly that the most central (highest) assumptions would be those about unobservables. In the reconditioned metaphor there is no particular specification that this should be so, nor that the points at the bottom of the pyramid should correspond to distinctively 'observational' expressions. As we now know, nearly all points in the pyramid will have some direct experiential content, in the sense that they will nearly all on occasion be used to make direct observation reports.

For what it is worth, there does in general seem to be some tendency for the points (expressions) at the bottom of such a theoretical structure to be 'more observational' than those higher up. But in so far as this is so, and for whatever reason, the diminished status of observational usages makes it of no great significance.

2

THE THEORETICAL CONTEXT
ACCOUNT OF MEANING

1 RELATIVISM GENERALIZED

The arguments of the last chapter raise fundamental questions about the objectivity of science.

Recall the difficulties that arose within the context of the double language model for any attempt to define 'theoretical' terms in observation language. I pointed out that these difficulties raised doubts about the possibility of 'theoretical' statements being known, and, even more fundamentally, made it unclear whether 'theoretical' terms referred to anything in reality. The most serious difficulty was that the relation of 'theoretical' expressions to observable indicators seemed to depend on which general assumptions involving those expressions were accepted. This implied that there was no possibility of any serious evidential assessment of those general assumptions themselves nor of any other statements depending on them. And I argued further that the complete lack of any content to the idea of such statements being right or wrong meant that the terms in them could not be held to have determinate referents.

Since then, however, we have found reason to suspect the initial distinction between 'theoretical' and 'observational' language. This might be thought to resolve our problems about 'theoretical' terms. If there is no substantial difference between talk about 'atoms' and 'viruses' and talk about 'tables' and 'chairs', then surely it follows that the former talk is as unproblematic as the latter. In a sense this does of course follow. But unfortunately a better way of describing the situation would be to say that talk about 'tables' and 'chairs' has turned out to be just as problematic as talk about 'atoms' and 'viruses'. That is, we now have reason to doubt that *any* statements can be known to be true, or indeed whether *any* terms can be held to have determinate referents.

The central theme of the second half of the last chapter was that which sensory stimuli prompt the application of any term depends in the last resort on the accepted theoretical assumptions involving that term ('theoretical' now meaning just 'general', rather than specifically 'unobservable'). But if this is right then the earlier doubts over statements about unobservables apply to all statements, and for essentially the same reasons. If the sensory associations of *any* term depend on which general assumptions containing it are accepted, then how are those general assumptions themselves to be assessed? It seems that anybody is perfectly entitled to accept whatever generalizations he likes, then moulding the content of his terms to fit. And so the epistemological and semantico-ontological doubts occasioned earlier by the apparent arbitrariness of judgements about unobservables now become entirely general. (As before, a possible explanation for the lack of philosophical awareness of these general doubts might be the consensualist presupposition that the requirements of scientific rationality will automatically ensure that everybody accepts the right generalizations. If there is never any possibility of serious disagreement on accepted generalizations then any adjustment of the sensory contents of our terms can happily be read as uncontentious corrections of previous error.)

2 KUHN AND FEYERABEND

It might seem absurd to suggest that even scientific decisions on apparently observable questions are at bottom arbitrary. If science is not objective, what is? Even so, in the last couple of decades a number of philosophers have developed accounts of science which come close to the surprising view that science is indeed irredeemably relative to arbitrary theoretical choices. Perhaps the most prominent amongst these have been T. S. Kuhn and Paul Feyerabend.

Central to the attack these two have made on objectivist orthodoxy is their account of the meaning of scientific terms. Both adopt what can be called a 'theoretical context' view of meaning: the meaning of any scientific term, including those used to make observation reports, depends on the surrounding context of scientific theory. (See in particular Kuhn [1962], Ch. 9; and Feyerabend [1965a].)

At first sight this might seem an implausible theory of meaning. But it is easy enough to see how it might be extrapolated from various of the arguments put forward in the last chapter. Both initially, in connection with terms for 'unobservables', and later, for all terms, including 'observational' ones, we discovered that in the last resort what seemed to determine how a scientific term was used were the general assumptions containing it. Even if on a particular occasion a scientific term is applied in more or less direct response to observational experience, such an application always seems answerable to the theoretical assumptions containing the term in question. But what is it to say this but to say that what constitutes a term's meaning *is* the set of general assumptions containing it? The theoretical context account merely makes this idea explicit.

Of Kuhn and Feyerabend it is the latter who has been the more explicit in his pronouncements on meaning. He has argued thoroughly and repeatedly against the orthodox view that a scientific term can have its meaning fixed by its association with sense experiences. He allows, as of course he must, that at any given time some terms will be such that appropriately trained human speakers will be directly prompted to apply them in response to certain sensory inputs. But, Feyerabend urges against the orthodox view, such psychological processes can no more give the words involved a *meaning* than can the physical processes involved give a meaning to the physical fact that the pointer of some scientific instrument is resting at a certain point on the dial. An instrument reading means something to us only by virtue of our general assumptions about how the entities being measured behave and interact with our instruments. Similarly with the words in a human observation report: it is only when they are interpreted through a surrounding context of scientific theory that they gain a significance. Understanding a scientific term is not a matter of being triggered to utter it by certain inputs but rather of knowing how that term is integrated into an accepted framework of general postulates. (Cf. Feyerabend [1962], pp. 35–42.)

The 'theoretical context' account of meaning encapsulates just those of the last chapter's conclusions which pose a threat to scientific objectivity. But, as I have observed, the threat involved is one that can remain unrecognized as long as it is taken for granted that there will always be unproblematic agreement on which new generalizations are to be accepted. However, there has been no danger of the relativist implications of Kuhn and Feyerabend's analysis of scientific language being overlooked. For both have views of scientific development in which the possibility of fundamental theoretical *dis*agreements plays a central part.

I shall consider first Kuhn's view of scientific development as outlined in his influential *The Structure of Scientific Revolutions* ([1962]). A central notion in Kuhn's account of science is that of a 'paradigm'; he holds that the scientists working in a given field at a given time will all share a single such paradigm. An important part of having a shared paradigm will be a common commitment to certain fundamental postulates. A paradigm will also contain certain metaphysical and methodological principles, plus examples of certain classic problem solutions to serve as exemplars for further research.

Kuhn takes the installation of such a paradigm in a scientific field to be a prerequisite of proper scientific research. The shared acceptance of a paradigm allows scientists to get down to 'normal science'. Normal science consists of refining and articulating an accepted paradigm. This will involve such activities as devising sophisticated experimental equipment, discovering more and more precise values for the physical constants entering into the paradigm's assumptions, finding satisfactory analyses of awkwardly complicated or apparently anomalous empirical phenomena, etc. It is only when there is agreement on basic assumptions and on what counts as successful research that scientists will have an adequate foundation for the detailed technical 'puzzle-solving' of normal science which enables them to develop sophisticated and precise theories. Disciplines in which no single paradigm has yet managed to establish itself are still in a 'pre-paradigm' stage: in such cases there will be nothing to stop continued contention about philosophical and metaphysical fundamentals preventing any development of detailed empirically applicable theories.

Once a paradigm has been established in some area it will normally remain there for some extended period. But according to Kuhn there will periodically be a scientific 'revolution', a traumatic episode during which an existing paradigm is abandoned and replaced by another. This will occur when the old paradigm has exhausted its fertility, when the few remaining puzzles arising within it prove resistant to solution. During the heyday of an established paradigm scientists working in some field will be entirely uninterested in alternative sets of basic assumptions. But at times of 'crisis', when the old paradigm seems to have run down, they will become susceptible to such alternatives; and when one offers itself there will in time be a general conversion heralding a new period of normal science.

The important point to note here is that the switch from an established paradigm to a new one is not a simple and uncontroversial matter of evidence showing clearly that the new ideas are to be preferred to the old. By the nature of the case this will be impossible. The adoption of a new paradigm will require that all previously accepted analyses of specific problems be reworked, for all those analyses would have been premissed on the now discarded assumptions of the old paradigm. In effect this reworking will be a major part of the task of the normal science to be conducted under the new paradigm. And as such it will take time and resources. What this means is that the adequacy of the new paradigm, its ability to deal with the empirical problems facing it as successfully as its predecessor did, will always have to be taken on trust at the time of the switch. Those discarding the old paradigm for the new will not yet be able to appeal to a greater range of problem solutions which uncontroversially validate their new allegiance. If at all, those solutions will become available only some time after they have committed themselves to the new paradigm, and only as a result of that commitment.

For example, when the heliocentric theory of the solar system was first proposed by Copernicus the predictions it yielded were in general not as accurate as those of its Ptolemaic predecessor. It was by no means clear that the Copernican system would be able to develop a corresponding predictive ability. And this is not even to mention the perceptual and dynamic problems occasioned by the counter-intuitive claim that the earth was in motion (cf. Feyerabend [1975], Chs. 6–8). Indeed the Copernican system won very few adherents until well into the seventeenth century: its merits became manifest only after the initially faithful had been working on it for many years

We can use the pyramid metaphor to explicate the point at issue. At the risk of greatly oversimplifying Kuhn's ideas, let us identify the adoption of a paradigm with a decision as to what form the most central, uppermost struts in our pyramid are going to take. Normal science is then the business of filling in the rest of the pyramid in a way that will adequately support just that crown of uppermost struts. Commitment to a paradigm is necessary if we are to have any idea of what kind of further building to do—if we don't know what we are going to support at the top we won't know what kind of further structure will be needed. But on the other hand there is no guarantee that once a given commitment has been made that it will be possible to build a supporting structure which is as strong and as widely based as, say, the supporting structures of our previous pyramids. It will only be after the commitment has been made and construction actually attempted that we will know if the new pyramid is an improvement.

Feyerabend agrees with Kuhn that an essential part of science is the adoption of fundamental assumptions which play a part in directing the analysis of specific situations. But the moral he draws from this is different from Kuhn's. For Kuhn the danger is that lack of agreement on a paradigm will prevent scientists from getting down to proper research. For Feyerabend on the other hand the danger is that an established orthodoxy will become dogmatically entrenched for no other reason than that it is the current orthodoxy. If the results which might show an alternative framework of fundamental ideas to be superior to the currently established view become available only after at least some researchers have committed themselves to that alternative, then might not theoretical unanimity simply perpetuate an inferior point of view? So Feyerabend urges theoretical proliferation—he recommends that scientists should continually strive to develop and pursue competing alternatives to established views, even when those alternatives appear to fly in the face of currently indubitable 'facts'. He thinks that it is only by encouraging such proliferation that we can prevent dogmatism, that we can stop an accepted view surviving beyond the point where it is valid for it to do so. (See in particular Feyerabend [1963].) As to Kuhn's worry that constant competition between fundamental ideas will merely lead to continual metaphysical disputation, Feyerabend suggests that this will be avoided effectively enough provided each scientist, or sub-group of the profession, is variously serious about his framework; there is no necessity for all to be committed to the same framework. (See Feyerabend [1970a], pp. 107–8.)

The most significant point for us is that both Kuhn and Feyerabend agree that significant advances in science involve the adoption of views the worth of which must at least initially be taken on trust. For this point of agreement implies that there is every possibility that different groups of scientists will on occasion quite legitimately diverge on which fundamental assumptions are to be accepted. Feyerabend thinks that this possibility is (or at least ought to be) continually actualized, while Kuhn takes it to happen only in pre-paradigm phases and, more relevantly, in the revolutionary interregnums between periods of normal science. But these differences are unimportant beside their basic agreement that such divergence can on occasion indeed take place.

As earlier observed, the relativist implications of a theoretical context account of meaning can be obscured by a consensual view of science. Conversely, if we still held to the view that the meanings of scientific terms are fixed by their experiential associations, then it would perhaps remain possible to view the promissory nature of much scientific development as simply being a matter of scientists having to wait to find out if a new theory is superior to an old one—just a matter of the time and resources required to show that the new assumptions are indeed consonant with the empirical data. But if meanings depend on theoretical contexts then the need to adopt fundamentally new ideas on trust does not just mean that scientists will have to be patient to find out if those new ideas are indeed improvements on the old—it seems to destroy any notion of one point of view being better than another. It is one thing for it to take time to find out whether one can build an adequate supporting structure for a certain type of pyramid tip. But if there are no constraints on where one can put foundation piles for the structure, if there is nothing to dictate how the constituent terms in a theory are to be used other than the shape of that theory itself, then it appears that any tip will in the end be as well supportable as any other. It would seem impossible for a scientist ever to be rationally compelled to admit that the views of his opponents are preferable to his own.

A number of critics have pointed out that Kuhn's and Feyerabend's views make an extreme relativism quite inescapable. And they have understandably found this relativism impossible to stomach. (See in particular Shapere [1966]; Scheffler [1967].) On the other hand it has proved difficult to produce any satisfactory response to the arguments of Kuhn and Feyerabend. The situation can be characterized as the 'paradox of meaning variance'. There do seem to be good arguments for moving away from a traditional empiricist conception of meaning towards a theoretical context account. But then it follows that with every change of theory there will be a change of meaning (the 'meaning variance thesis'). And with this there seems no way of escaping the unpalatable consequences that objective choices between scientific theories are impossible.

In their earlier writings neither Kuhn nor Feyerabend apparently saw any great tension between their general vision of scientific progress and the theoretical context account of meaning. They independently coined the term 'incommensurability' to characterize the situation where two competing theories each gave different meanings to all the terms used in the relevant field (cf. Feyerabend [1970b], p. 219). In a sense 'incommensurable' theories were about different worlds—they pictured the world as containing different kinds of things, even down to the level of observable entities. But the difficulties that 'incommensurability' placed in the way of rational choices between theories seemed originally to be taken merely to add weight to the idea that switches from one theory to another could not in the first instance be rationally compelled but had to depend initially on faith and commitment.

Kuhn and Feyerabend have responded rather differently to their critics' insistence that if incommensurability is taken seriously scientific theory-choice is not just a matter of initial faith and commitment—it is never anything but faith and commitment. Kuhn has come to emphasize the possibility that there are after all certain impartial bases of comparison with respect to which some theories can be shown to be objectively better than others. (Kuhn [1970a], [1970b].) Feyerabend on the other hand has simply embraced an explicit relativism. Theoretical proliferation is no longer to be defended because it leads to better theories later, but, if at all, simply because it is in itself intellectually liberating and aesthetically pleasing. In the end, urges Feyerabend, 'anything goes' in science. (Feyerabend [1970a], [1975].)

3 THEORETICAL CONTEXTS

Feyerabend has never been entirely specific on what exactly is to be understood by 'theoretical context'. Clearly the theoretical context of a term is something to do with the set of universal generalizations containing it and related terms. But are all such general assumptions to be counted as part of a term's theoretical context? Or only some? Or what?

It would make a kind of initial sense to count all the *lawlike* generalizations containing the terms used in a certain area as constituting their theoretical context. What distinguishes lawlike generalizations from merely accidental ones is their ability to sustain counterfactual and subjunctive conditionals. For example, the lawlike generalization 'All gases expand when heated at constant pressure' implies 'That sample of gas would expand if it were heated at constant pressure.' But the accidental 'All the coins in my pocket are silver' does not sustain 'That penny would be silver if it were in my pocket.'

J. L. Mackie ([1962]) has explained the differences between lawlike and accidental generalizations in terms of the different kinds of evidential grounds respectively required to warrant their acceptance. Lawlike generalizations can legitimately be accepted prior to an exhaustive examination of all their instances. Thus we are entitled to accept 'All gases expand when heated at constant pressure' even though we have not, *per impossibile*, examined all instances of gases heated at constant pressure. By contrast, someone would not be entitled to conclude 'All the coins in my pocket are silver' unless he had in some way checked individually on each of them. This contrast then accounts for the differential ability of lawlike and accidental generalizations to sustain counterfactuals: if an accidental generalization can only be accepted after an exhaustive examination of their instances, then the 'introduction' of a hypothetical and therefore unexamined instance will, so to speak, remove our original warrant for accepting the generalization and so prevent us from using that generalization as a premiss in any reasoning about that hypothetical instance. The acceptance of a lawlike generalization on the other hand does not presuppose that all its instances have been examined and found to conform, and so within limits we will be able to continue using the law in our reasoning about hypothetical further instances.

Of course this story in a sense only pushes the problem of distinguishing between accidental and lawlike generalizations back a little way. For it does nothing to explain why it is that some regularities are legitimately 'projectible' to as yet unexamined instances while others are not. But what it does do is show why someone might be attracted to the view that the meanings of scientific terms depend on all the lawlike generalizations accepted in the relevant field; that is, to the view that the 'theoretical context' which fixes the meanings of terms in any field is constituted by the totality of accepted lawlike generalizations. For there is a sense in which accepting a lawlike generalization gives us a new criterion for applying the linguistic terms it contains. Consider the very simplest case, a generalization of the form $(x)(Ax \supset Bx)$. In accepting this as a law we commit ourselves to the view that as yet unexamined As will be Bs (and that all as yet unexamined not-Bs will be not-As). And so we will be led to apply the term B to any individual that we discover to satisfy A (and conversely to apply not-A to any individual that we discover to satisfy not-B).

Before examining the suggested interpretation for 'theoretical context' rather more seriously a couple of peripheral observations will be helpful. Firstly, it is worth noting that not all the lawlike generalizations accepted in a given scientific community will be explicitly articulated. For example, the assumption, central to classical mechanics, that the mass of a body is independent of its velocity literally went without saying. There is clearly no reason to exclude unarticulated generalizations from a 'theoretical context'. The argument is in effect that lawlike generalizations provide an inferential procedure governing the use of the terms involved. A generalization will do this no less for being unarticulated. Newtonians certainly took information about the mass of a body at one velocity to be an adequate basis for inferring it had the same mass at other velocities.

Secondly, there is the point that in their most perspicuous form scientific generalizations will almost invariably contain no non-logical expressions other than predicates. But scientific discourse also essentially involves singular terms: in particular, singular terms purporting to refer to spatio-temporal particulars are indispensable for any description of the results of experiments and other particular occurrences. So it seems that a 'theoretical context' account of the meanings of scientific terms must be seriously incomplete. However this is scarcely a major problem. In the present context we can take it that the singular terms for spatio-temporal particulars essential to science function by conjoining one or more predicate terms with some kind of demonstrative convention (such as govern particles like 'this', 'here', 'now', 'there', 'then', and as are involved in certain uses of the definite article). And so, presuming that such demonstrative conventions are themselves unproblematic, we can take the 'theoretical context' account to deal with the relevant singular terms via what it says about the predicate expressions they are constructed from.

So the suggestion to be examined is that the meanings of scientific terms are given by the 'theoretical contexts' constituted by all the generalizations accepted in the relevant field. (It will simplify the exposition if from now on all generalizations are taken to be lawlike and all scientific terms are taken to be predicates, unless otherwise specified.) The obvious objection to this suggestion is that it implies that any change in any accepted generalization will change the meanings of all the terms used in that field. Any two sets of generalizations which differ in even the slightest respect, will come out as 'incommensurable' theories, with all the attendant difficulties for objective scientific decisions. Any decision to reject any generalizations for another will simply be a matter of deciding to use words with different meanings rather than responding rationally to agreed evidence. And so on the present suggestion we have a breakdown of rationality not only when fundamental assumptions about, say, the nature of 'forces' are at issue. We have the far less plausible thesis that this breakdown will occur with even the slightest theoretical modification, such as the alteration by some small amount of the constant G in the Newtonian law of universal gravitation. Surely this cannot be right. It seems absurd to deny that scientists working within classical mechanics can have rational evidence for concluding that the value they previously attached to G was imprecise.

Can this really be what Feyerabend has in mind when he argues that meaning depends on theoretical context? As mentioned in the last section, Feyerabend wants to prevent accepted scientific theories commanding continued acceptance simply because they happen to be the established orthodoxy. He says in one paper that his ultimate intention is to create 'an abstract model for the acquisition of knowledge . . . which has as its *aim* maximum testability of our knowledge'. ([1965b], p. 223.) It would be surprising indeed if somebody with such Popperian ideals were committed to a theory of meaning which implied that nothing could ever rationally compel any scientist to revise any of his theoretical assumptions

However there is little, at least in Feyerabend's early writings, to give us any reason to think that this is not indeed part of his theory of meaning. Thus in both his [1962] and his [1963] Feyerabend argues that it is impossible to define in Newtonian mechanics a term with the same meaning as the 'impetus' of late medieval Aristotelian physics. His reason is that:

The concept of impetus is . . . formed in accordance with a law (forces determine velocities) and this law is inconsistent with Newton's theory and must be abandoned as soon as the latter is adopted. ([1962], p. 34.)

Feyerabend does not, in this typical argument, produce any grounds for thinking that there is anything special about the law in question that makes it constitutive of the meanings of terms related to it.

4 ANALYTIC VERSUS SYNTHETIC

A number of writers have pressed Feyerabend on whether he really wants to say that *all* theory changes, however slight, are to count as rationality-disrupting meaning changes (e.g. Shapere [1966], pp. 53–6).

In response to this kind of pressure Feyerabend has in more recent articles made a number of brief suggestions as to how the theoretical context account might be saved from this conclusion. Thus in his [1965b] he maintains that the postulates, or rules, of a theory

> . . . form a hierarchy in the sense that some rules presuppose others without being presupposed by them. A rule R' will be regarded as more fundamental than another rule R", if it is presupposed by more rules of the theory, R" included, each of them being at least as fundamental as the rules presupposing R". . . . Changes of fundamental laws are regarded as affecting meanings while changes in the upper layers of our theories are regarded as affecting beliefs also. (p. 259.)

Later, in Section 5 of Chapter 4, I shall develop some ideas similar in spirit to what Feyerabend seems to have in mind in this brief suggestion (in fact made in a footnote). But I shall not argue that they solve Feyerabend's problem. For there remains the question, which Feyerabend does not answer, of how a comparative ordering of rules into 'more' and 'less' fundamental can lead to an absolute distinction between fundamental and other laws, such that changes in the former but not in the latter alter meaning. It would still seem that any change in laws changes meanings, albeit perhaps to different degrees. In his [1965c] Feyerabend considers the rules governing scientific terms as grouping particulars into classes. He then argues that there will be stability of meaning when a theoretical innovation does no more than produce membership changes within that system of classes. Meaning changes occur when theoretical innovations 'change the system of classes itself'. Again, the idea that scientific terms group particulars into classes is important and worth pursuing, as I shall do in some detail in the next chapter. But there too we shall run into a problem which Feyerabend does virtually nothing to deal with: what distinguishes amongst theoretical innovations those which merely rearrange the assignment of particulars to the existing system of classes and those which produce changes in the system of classes itself? In his [1975] Feyerabend suggests that it is when 'covert classifications' depending on 'hidden ideas' or implicit 'universal principles' are 'suspended' that meaning changes occur and incommensurability arises (pp. 224–5, 269–70). This is in a way a more definite suggestion than Feyerabend's earlier attempts to modify the theoretical context account. But it is difficult to see why we should accept it. I noted in the last section that there was as much reason to suppose that implicit unarticulated postulates fixed meanings as to suppose that explicit postulates did so. But we are now being asked to suppose that *only* implicit postulates fix meanings. It is not at all clear why the articulation of some assumption should in itself disqualify it from this function.

Feyerabend's inability to come up with a satisfactory solution here is not exactly surprising. For, despite his occasional protestations to the contrary (e.g. his [1965b], p. 259), it is clear enough that what he wants is some way of separating the postulates in a theory into *analytic* ones and *synthetic* ones. For what he is after is some way of dividing the generalizations in a theory into those which can only be revised by changing meanings and those which can get revised because of new evidence—into those whose acceptability flows immediately from the meanings of their constituent expressions and those whose acceptability depends in addition on evidence about the way the world is. What else would such a division amount to but an analytic-synthetic distinction?

(It might be objected that Feyerabend's intended distinction differs from the analytic-synthetic distinction in that he clearly thinks that changes in meaning-constituting postulates can have substantial import, that they can amount to new claims about the way reality is. Whereas on the traditional conception changes in 'factually empty' analyticities could only signify terminological shifts, or at best a decision to change the subject. However it is by no means clear at this stage of the argument that Feyerabend is *entitled* to think that semantic variation can be relevantly non-trivial.)

It is clear enough how an analytic-synthetic distinction would help with the 'paradox of meaning variance' and related problems. If scientific theories did split up neatly into analytic and synthetic postulates there would be no particular difficulty about the ability of contrary evidence to compel the revision of lawlike generalizations. For the analytic postulates in a theory would fix the meanings of the terms involved in such a way that, when new evidence arose, it would be objectively indicated which, if any, of the synthetic postulates needed revising. More generally, competing points of view could always be decided between by recourse to agreed evidence, provided only that the dispute was restricted to what synthetic generalizations should be accepted, and was based on an agreed core of analytic postulates fixing shared meanings for the terms used in the dispute. Indeed, as might have occurred to some readers, an analytic-synthetic distinction could well have helped with some of the problems raised early on in the last chapter. In particular, one way of blocking the relativistic difficulties about 'theoretical' statements in the double language model would have been to distinguish analytic correspondence rules from synthetic correspondence rules. Such a distinction need not have amounted to a definition of 'theoretical' terms in 'observational' language, but it would nevertheless have offered some account of what stopped scientists simply revising any correspondence rules that did not fit in with their favoured theoretical postulates.

The trouble with all this (and the reason I did not introduce the notion earlier) is of course that there are serious objections to the idea of an analytic-synthetic distinction. There is an extensive modern literature on this topic, beginning in effect with Quine's [1951]. The issues involved interact with the overall theme of this book at a number of points, and it would be impossible to deal with them all immediately. I shall use the rest of this section for a preliminary treatment; and shall then return to the topic at a number of subsequent points.

The basic Quinean objection is that it is not at all clear what substance there is to the supposed distinction. In the context of our problem, for instance, it is clear enough that scientists accept structures of (explicit and implicit) postulates of generalized form containing the various expressions they use in their field. But what reason is there to make the further supposition that these structures have two quite different kinds of components—some meaning-fixing and inviolable and others substantive and corrigible? There is no obvious behavioural or textual evidence that scientists themselves discriminate between their postulates in this way.

There is a natural temptation to argue that if in a given scientific development certain generalizations are revised, while others remain inviolate, this in itself shows that the former must have been synthetic, while (some amongst) the latter were analytic. That is, it is tempting to take the facts of scientific development themselves to show us where the distinction between analytic and synthetic lies. But in the present context of discussion this ploy is quite inadmissible. If we are concerned to find features of scientific language and practice which will explain how it is possible for scientists to revise and decide between theories objectively, it will not do to read back from the actual development of scientific theories to the dividing line between analytic and synthetic. For this will scarcely then give us an account of *why* the theories were developed in that way, of what guided scientists to revise their theories in that way rather than in others. The difficulty here is not just that we are no longer entitled to presuppose the objectivity of scientific development. Strictly speaking, of course, the paradox of meaning variance calls even this in question. But even if we stand by our objectivist intuitions and take it for granted that science does somehow proceed objectively, our task is still to give some explanation of how this is done. An 'analytic-synthetic' distinction cannot play a real part in this if the only thing that makes some sentences synthetic and others analytic is that the former are the ones that do, as it happens, get revised. If the analytic-synthetic distinction is to be of serious assistance here, if it is to yield the substance rather than the mere form of an explanation, then clearly we need to find some further difference between analytic and synthetic statements.

The dangers in an empty and unargued appeal to the analytic-synthetic distinction can be illustrated by reference to some of the arguments in Adolf Grünbaum's 'Can We Ascertain the Falsity of a Scientific Hypothesis?'. (Grünbaum [1971]. See also Grünbaum [1960], and the comments on Grünbaum in Harding [1976].) Grünbaum is in this paper concerned to dispute the view, which he attributes to Pierre Duhem, that it is never possible for observational evidence to show effectively that any specific constituent hypothesis in a scientific theory is false. He maintains that this thesis can be upheld only if we allow hypotheses to be saved by trivial re-adjustments of their constituent terms' meanings. Once such trivial semantic ploys are ruled out, then, according to Grünbaum, it is perfectly possible to produce counter-examples to the 'Duhem thesis'.

Clearly this argument obliges Grünbaum to say something about what distinguishes episodes of semantic stability from those which do involve meanings changes. For this he turns to Peter Achinstein's [1968]. Achinstein suggests we can distinguish between 'semantically relevant' properties for something's being an X, and 'non-semantically relevant' such properties. Semantically relevant properties are those which 'in and of themselves' count in favour of (positive relevance), or against (negative relevance), classifying something as an X; while possession of a non-semantically relevant property might do so 'solely because it allows one to infer that the item possesses properties of the former [semantically relevant] sort' (pp. 8–9). Later Achinstein adds that:

> Properties semantically relevant for X . . . have a particularly intimate connection with the meaning of the term "X" . . . as other properties do not. . . . It is perfectly possible that there be two different theories in which the term "X" is used, where the same set of semantically relevant properties of X . . . are presupposed in each theory. . . . If so, the term "X" would not mean something different in each theory. (p. 101.)

Unfortunately Achinstein lacks any conclusive arguments to back up his belief in the supposed distinction between semantically and non-semantically relevant connections between properties, and much of the burden of filling it in has to be borne by putatively illustrative examples.

Let us now examine Grünbaum's application of Achinstein's supposed distinction. Grünbaum considers amongst other episodes the discussion generated by the Michelson-Morley experiment. He maintains that the terms of the discussion make it clear that 'both parties to the dispute agreed that, to within a certain numerical accuracy, equal numerical verdicts furnished by rigid rods were positively semantically relevant to the geometrical relation term "spatially congruent" (as applied to line segments)' (p. 103), whereas the (negative) 'relevance of the equality of the round-trip times of light to *spatial* congruence was made *contingent* on the particular *law of optics* which would be borne out by the Michelson-Morley experiment' (p. 104). That is, Grünbaum maintains that equal numerical findings by rigid rods (RR) were semantically relevant to spatial congruence (X), but that being traversed by light in equal round-trip times (Y) was only non-semantically relevant.

However Grünbaum's basis for this claim is in the end simply the fact that 'standard treatises' account for the Michelson-Morley experiment by rejecting the (classical) ether theory's assumption that light generally takes different times to traverse equal spatial intervals, and replacing it by the (special relativistic) tenet that the transit times in such cases will always be equal—that is, the fact that scientists decided to stick with the original assumption about the relation between RR and X, and to change the postulated connection between Y and X.

But, as observed in general above, this is no way to draw an analytic-synthetic distinction (or a distinction between semantically relevant and non-semantically relevant properties) if the distinction is then supposed to explain how it is that observational evidence rationally compels that certain existing hypotheses rather than others should be deemed false. If Grünbaum hopes here to show us what makes it possible for us to falsify a hypothesis, that we *can* falsify a hypothesis (if indeed that is the correct way to describe the development), then he surely needs to produce some further distinction between the Y–X relation and the RR–X relation other than that the first was in the end abandoned and the second was not. (He does, it is true, also argue that according to relativity physics RR, unlike Y, bears the same relation to X on rotating discs and in other non-inertial frames of reference. But he does not explain why this kind of invariance should have anything to do with semantic relevance.)

In the particular case at hand the question-begging nature of Grünbaum's analysis is made manifest by the well-known fact that by no means all scientists at the time did immediately reject the existing classical hypothesis connecting Y and X. For there was the alternative analysis of the Michelson-Morley experiment proposed independently by Lorentz and Fitzgerald in the 1890s. This weighs positively against Grünbaum's claim about 'usage . . . at the turn of the current century' (p. 104). For what Lorentz and Fitzgerald did was to deny the connection between RR and X which Grunbaum diagnoses as 'semantically relevant'. Their suggestion was that rigid bodies contracted in the direction of their absolute motion by a factor $\sqrt{1 - \frac{v^2}{c^2}}$ (where v is their absolute velocity and c the absolute velocity of light). In effect they opted to stick with the existing classical hypothesis connecting Y and X (light would take different times to traverse equal intervals with different absolute motions) and to adjust the supposed relation between RR and X instead. If anything their strategy would support the claim that it was Y and not RR that was semantically relevant to X 'at the turn of the current century'.

It might seem open to Grünbaum to argue here that the Lorentz-Fitzgerald contraction changed the meanings of the terms used in the classical ether theory, whereas special relativity did not. But it should by now be clear that Grünbaum has got no way of supporting this thesis. Lorentz and Fitzgerald were perfectly serious scientists, and however tempting it may be, we have been given no warrant for accusing them and their followers of simply playing with the meanings of words. Indeed, if one had to choose, it would be far more plausible to say that it was special relativity that was guilty of creating semantic instability. If anything it was surely Einstein's theoretical innovations rather than Lorentz's and Fitzgerald's that did more damage to established linguistic usages.

There is a more general moral to be drawn from this story. If we are to be able to show that any theory-choices in science resulted from objective considerations, then presumably we ought to be able to do this for the particular case of the eventual victory of Einstein's modification of classical mechanics over the Lorentz-Fitzgerald alternative. Surely Einstein's formulation of relativity physics was a case of a scientific advance, if anything is. The above discussion suggests that whatever did make Einstein's response the better one it is unlikely to have been a mere matter of its conforming to semantic constraints which were violated by the alternative. For, as just observed, it is far more plausible to construe the Lorentz-Fitzgerald suggestion as maintaining semantic stability than to suppose that relativity theory did so. (For an account of what eventually showed the superiority of Einstein's proposals in terms of the kind of methodological considerations discussed in Chapter 4 below see Zahar [1973].)

But this is all jumping the gun somewhat. For at this point we are not yet strictly entitled to suppose that *any* scientific theory-choices are objective—we have as yet done nothing to dispel the fundamental doubts about scientific objectivity raised earlier by the meaning variance thesis.

Let me sum up the situation. The arguments of Chapter 1 seemed to lend support to Feyerabend's contention that the meanings of all scientific terms depend on their theoretical context. But this dictum seemed to imply that objective choices between competing scientific points of view were impossible. So we then asked what exactly a 'theoretical context' was supposed to be. There seemed some sense in counting *all* the lawlike generalizations accepted in a given field as constituting the theoretical context of terms in that field. But of course this only made things worse—for it then seemed that not even the slightest alteration of any accepted generalization could be rationally compelled by objective evidence. Perhaps then only *some* accepted postulates (the analytic ones) fixed meanings? But, so far at least, we have been unable to attach any substance to this latter thought.

5 OBSERVATION REVISITED

We seem to have reached something of an impasse. In the face of these difficulties perhaps we ought to take another look at the connection between observation and meaning. The initial argument for the theoretical context account of meaning was that theoretical assumptions seemed to dominate over any associations scientific terms may have with sensory experience. But there remain reasons for not entirely discounting the relevance of experiential associations to meanings. For one thing, note that the use of scientific predicates cannot be governed solely by inferential rules sustained by theoretical postulates containing them. For by themselves such rules of inference could never suffice for any application of any predicate to a specified particular. That B can justifiably be asserted of any particular to which A is applicable will only show that any particular actually is a B if we already have reason to judge it an A. And we could only *infer* it's A-ness if we already had reason to ascribe other properties to it. It is clear enough that if this kind of thing is ever to get started there will have to be some *non*-inferential ways of applying some predicates to particulars. What fills this gap is obviously direct observational applications of terms. More pertinently, if it is ever to be possible for evidence to require the revision of a lawlike generalization, then surely it can only be as a more or less direct result of such observational applications of predicates. The standard reason for giving up a generalization is the acknowledgement that a particular situation provides a counter-example. And, as just explained, any singular statement characterizing such a situation has to stem in the end from some observational application of some predicate. In the last section but one I took it that the revision of generalizations could at least sometimes be rationally compelled by evidence. We now see explicitly that what this means is that the theoretical assumptions involving scientific terms do not *always* dominate over their experiential associations in cases of conflict.

But there are difficulties in the way of reinstating experiential associations as influences on the meanings of scientific terms, precisely analogous to those facing the theoretical context account of meaning. Suppose we conceive of the observational use of a term in the way agreed at the end of the last chapter, as being governed by a connection between that term and certain physiological stimuli. I shall speak (perhaps misleadingly, but not seriously so) of such connections as giving rise to 'observational procedures'.

Now, are all observational procedures to affect meanings? The arguments of Chapter 1 have already given us reason to suppose this cannot be right. For it was there argued that it is perfectly possible on occasion for our theoretical knowledge to show us that certain observational procedures are unreliable and in need of revision. But if a given observational procedure can be so discovered to lead us astray, then surely *it* at least cannot play any part in fixing the meanings of the terms it involves.

Perhaps then not all observational procedures influence meanings, but only some special sub-set thereof? Even if some observational procedures are contingent and revisable, might there not be a core which are not, perhaps those relating to our most natural and basic observational abilities?

There is an obvious resemblance between this last suggestion and the idea of an analytic-synthetic distinction amongst theoretical postulates. And unfortunately just the same objections apply. In the first place it is not at all clear what is to give substance to the suggested distinction between contingent and revisable observational procedures and basic meaning-fixing ones. Scientists certainly use sets of observational procedures; but they do not explicitly distinguish amongst those procedures in the required way. Nor is there any other obvious feature by which we might pick out the 'analytic' observational procedures. Moreover, by any intuitive standards it seems that some of the examples of actually rejected observational procedures discussed in the last chapter would have to be counted as basic ('red', 'stationary'). In the end there seems little reason to suppose that there is in fact any privileged category of observational procedures which are in principle immune from revision. (For more detailed argument on this point see Hesse [1974], Ch. 1, Sec. II.)

There is of course always the further circumstance that as it happens certain observational procedures do get revised and others do not. But as shown in the last section this is a quite inadequate basis for the kind of distinction we are after—if we are trying to explain what enables science to proceed as it does we need to have at least an abstract idea of what *influences* the developments that occur. A distinction whose only substance is derived from which developments *do* occur is not good enough.

So now we have reached the same impasse with observational procedures as was reached with theoretical generalizations. It seems clear that all observational procedures cannot fix meanings. But on the other hand we seem unable to give any good sense to the idea that only some such procedures do so.

Whichever way we turn, it seems impossible to come up with any satisfactory account of the meanings of scientific terms. How then are we to evaluate the 'meaning variance thesis' and its paradoxical implications? In the next chapter I shall start by considering a line of approach which promises a way of coming to grips with these issues even in the absence of a satisfactory theory of meaning for scientific terms.

3

SENSE AND REFERENCE

1 SENSE VERSUS REFERENCE

A number of recent philosophers argue that the 'paradox of meaning variance' arises only when insufficient attention gets paid to the distinction between *sense* and *reference*.

This distinction was first made properly clear by Frege. He showed how in the theory of language we need to separate the independent existent that a word stands for (its 'reference') from the aspect of that word's use that determines it to have that reference (its 'sense'). In a manner of speaking, we can conceive of the sense of a word as that which directs us to the thing or things that it stands for. (Frege [1892].)

As Frege also made clear, with the well-known example of 'Morning Star' and 'Evening Star', it is perfectly possible for two words with different senses to have the same reference: 'names with different senses but the same reference correspond to different routes leading to the same destination' (Dummett [1973], p. 96).

It is this latter point that has been appealed to in connection with the meaning variance thesis and the associated problems of relativism. Thus Israel Scheffler, in his [1967], admits that it is plausible enough to suppose that the *senses* of scientific terms might depend somehow on their theoretical contexts, and so might vary with theoretical changes. But, he points out, it by no means follows that the *references* of scientific terms will vary with theoretical changes. A term can have different senses in different theories yet still be used to stand for the same thing.

In the last chapter our failure to come to proper grips with the meaning variance thesis was due to the lack of an adequate account of the 'meanings' of scientific terms. But we were there considering scientific terms almost exclusively from the point of view of *sense*, from the point of view of what guided the use of scientific terms in application to reality. Scheffler's suggestion is thus in effect that we should bypass the obstacles which blocked progress in the last chapter, by simply shifting our attention from sense to reference.

In support of his recommendation Scheffler points out that it is concepts of reference, rather than of sense, which are central to our understanding of such basic notions as *truth*, and *logical implication* and *incompatibility*. Suppose we call what a singular term stands for its *denotation*, and the class of things a predicate stands for its *extension*. Then we can specify that a singular statement of the form F*a* is true just in case the denotation of *a* is included in the extension of F. Again, the truth of a generalization of the form (x) (Fx ⊃ Gx) requires just that F's extension be contained in G's. Following Tarski, we can go on from this kind of basis to give a recursive specification of what it is for more complex sentences to be true in terms of the referential values of their constituent expressions. Correspondingly, logical relations between different sentences depend on the references of their constituent expressions rather than on those expressions' senses. For instance two sentences are logically incompatible if the referential values of their constituent expressions are related in such a way that the two sentences cannot both be true. An obvious and important particular case is where two sentences respectively attach a predicate and the negation of that predicate to some singular term.

So Scheffler argues that the possibility of sense inevitably varying with scientific theories is unimportant: 'Whether synonymies are sustained or violated is, in itself, . . . of no real interest for the general purposes of science.' In particular he argues that as long as terms in different theories *refer* to the same things then 'the subjectivist argument from meaning collapses' and we have reinstated the possibility of 'disagreement in the sense of explicit contradiction, . . . itself involved in any plausible conception of rational discussion'. Again, the possibility of constancy of 'referential identities throughout variations of theoretical context' lends substance 'to the notion of control or justification of theory' (Scheffler [1967], pp. 60–2).

(Variations on Scheffler's approach are to be found in articles by Michael Martin [1971] and Arthur Fine [1967]. Martin argues that the possibility of objective decisions between scientific theories will be assured not only when they have terms with identical extensions but in any situation where certain relations of inclusion obtain between the extensions of their respective terms. And Fine maintains that statements of different theories will be sufficiently logically related for objective comparison whenever they contain expressions whose extensions coincide for at least some restricted range of objects. I shall not discuss these refinements as such; but the criticisms I make of Scheffler's arguments could easily enough be adapted to apply to these further suggestions.)

Perhaps Scheffler is right to insist that it is reference rather than sense that is of primary importance for the analysis of science. But there remains a serious hiatus in his approach. For it is clear that an appeal to referential notions will be an effective response to the paradox of meaning variance only in so far as it can be shown that the referential values of scientific terms do not themselves inevitably vary with scientific change. Scheffler says:

> The insistence . . . that theoretical incorporation affects meanings is plausible at best only with respect to senses, and even so only for certain theoretical incorporations. . . . Such alteration, however, . . . does not automatically effect a disruption of referential constancy nor, therefore, does it automatically disturb the deductive relations which underlie reduction and explanation. (p. 62.)

But he does not produce any substantial arguments for this confidence in the referential constancy of scientific terms. It is one thing to point to the 'Evening' and 'Morning' stars to show that terms with different senses *can* have the same reference. But this is scarcely the norm. References do, after all, depend on senses, and it is only in what are in a sense accidentally special cases that terms with different senses will fail to have different references. It seems quite open for someone of relativist inclinations to maintain that the referential values of scientific terms are as theory-dependent as their senses, varying whenever the theory in which they are used is modified.

Scheffler does, it is true, suggest that for some terms at least it will be possible for their referential identities to be determined 'independently of a characterization of their respective senses' by 'reinforcement through shared processes of agreement on particular cases', or by 'the possibility of shared processes of decision on the referential force of a term by application to particular cases' (pp. 61–4). But it is hard to see how this is supposed to work. In general scientific predicates will have essentially 'unbounded' referential values: there will always be new and as yet unexamined instances in the spatial and temporal distance for them to apply to. And this makes it difficult to see how their referential values can be fixed by 'agreement on particular cases'—we obviously cannot fix the referential value of such a scientific term by simply laying down that it applies to each and every one of a given number of identified individuals. (If Scheffler's emphasis is supposed to be on the 'shared *processes* of agreement', rather than the particular cases agreed on, then it remains to be shown, in a way Scheffler does not, that such shared processes in general survive theoretical change.)

2 EXTENSIONS

If we are to evaluate seriously the referential approach to the paradox of meaning variance we shall need to know what exactly it is for scientific terms to have certain individuals or classes as their referential values—to know what, if anything, makes it the case that they stand for those individuals or classes instead of others. It is only when we have answered this question that we will be in any position to tell what, if anything, can ensure that terms from different theories have the same references.

Unfortunately it will turn out that questions about the references of scientific terms are as awkward as questions about their meanings. (Henceforth I shall use 'meaning' specifically for sense rather than reference.) In particular we shall find initial reason in this chapter for doubting whether scientific terms do have determinate referential values at all (which will *a fortiori* be reason for doubting that terms from different theories can have the same references). And even though I shall return to the question of the referential values of scientific terms in later chapters, and in Chapter 5 in particular, we shall in the end find no sound basis for eliminating such doubts.

In what follows I assume we are dealing with a first-order language in which the singular terms and quantifiers are meant to range over spatio-temporal particulars. Actual scientific languages are of greater expressive power in various necessary ways. But it will greatly aid the exposition to make this simplifying assumption, and the kind of difficulty I shall raise is easily enough generalizable to more complicated cases.

Furthermore I shall concentrate on the notion of the extension of a predicate expression, assuming without argument that there are no problems about the denotations of singular terms and the ranges of our variables. That is, I shall assume that we have a well-defined universe of discourse of spatio-temporal particulars, and that certain of these particulars are variously the denotations of our singular terms. Since I am arguing at this point primarily *against* the advocates of a referential approach to our problems, I shall be happy to show that even granted this latter assumption there remain difficulties with the supposition that scientific predicates have well-defined sub-sets of that universe of discourse as their extensions.

It might seem unclear what difficulties there could be here. What could be more straightforward than the idea that scientific predicates stand for certain classes of individuals? And if some explication of the 'standing for' relation is really needed, can we not get it simply enough by inverting the Tarskian definition of truth for singular sentences and abstracting from the role played by singular terms. That is, can we not simply take it that any predicate stands for that class containing all those individuals it is 'true of', all those individuals that 'satisfy' it? However, this 'explication' only takes us round a fairly small circle, given that the sole account so far suggested of what it is for a predicate to be 'true of' an individual, for that individual to 'satisfy' the predicate, is for the individual to be in the extension of that predicate. There is, of course, a connection between talk of predicate's extensions and talk of the truth of sentences containing them. But all this means is that any difficulties about the former will be difficulties for the latter too.

Still, why should we need an explicit account of what it is for a predicate to have a given extension before we can start talking about predicate's extensions? Explication has to stop at some point. Surely it is enough that the extension of a given predicate P in a language can always be characterized directly in any metalanguage containing a name of P? Can we not always use some metalinguistic predicate Q to specify straightforwardly 'The extension of the object language predicate P includes all and only those things that are Q's'? And does not the possibility of such specifications in itself show that we can always legitimately talk of a given predicate's extension?

But the problem being raised cannot be dismissed thus simply. It is important to remember that we are not in the position of the formal logician who thinks of himself as 'interpreting' a formal object language. In a sense the logician is in a privileged position: he can specify whatever extensions he likes for the predicates in his object language, for his conception is of an object language which is strictly speaking an uninterpreted structure of meaningless symbols. But any metalinguistic characterization we give for a scientific language will need to be justified, for we are dealing with scientific languages which have a life of their own outside their metalinguistic 'interpretation'. For instance, the kind of metalinguistic specification illustrated above will be wrong unless Q in the metalanguage actually does have the same extension as P in the object language. So we can scarcely take such specifications for granted in a context of argument where the conditions for and the possibility of such coextensionality is precisely what is in question.

There is of course always the possibility of using some expansion of the object language as a metalanguage for itself. Surely this device will suffice to guarantee that our metalinguistic predicate and our object language predicate are coextensional—for will they not simply be the same predicate? But note that the possibility of this procedure casts no light on our initial problem of what it is for predicates in different languages to have the same extension. For given different object languages any metalinguistic interpretation will now have to be in different metalanguages, and there will remain the question of what it is for predicates in those languages to be coextensional. (Cf. Putnam [1973b], p. 201.) Nor, more generally, does the possibility of the suggested bootstrap operation mean that we can take the notion of a predicate's extension for granted, that we are entitled to suppose that any given predicate will be attached to a well-defined class of individuals as its extension. For if there *are* any difficulties in this respect they will scarcely disappear simply because we decide to consider the predicate as part of a metalanguage.

The possibility of using a metalinguistic predicate to specify the extension of some predicate in an object language does in a sense show how we can introduce talk of extensions directly without any further explication. But it does not thereby show that such talk is unproblematic. As the lesson of naïve set theory showed us long ago, the possibility of forming names which purport to stand for classes abstracted from predicates does not in itself guarantee that such abstractions will be successful.

So far I have obviously done nothing to show that there actually is anything suspect in talk of predicates' extensions. What I have attempted to show though is that the standard introductions to the notion leave it open that there might be. I shall now try to produce some positive grounds for suspicion.

Up to this point I have been ignoring the important and obvious point that in so far as predicates have extensions this must somehow depend on features of the linguistic practice of the community of human beings who use those predicates. Irrelevant onomatopoeic exceptions aside, it is clear that any relation between a predicate and the individuals satisfying it cannot derive simply from an inherent connection between the word and the things, but must emerge somehow from the way the relevant community conventionally uses the predicate.

What then are the relevant aspects of the linguistic practice of scientific communities? One point which needs to be remembered is the essential 'openness' of the range of individuals satisfying most scientific predicates: the possible candidates always stretch into the spatial and temporal distance. The way linguistic practice fixes predicates' extensions must be consistent with this openness. Thus, as we saw earlier, it cannot in general be that it is simply laid down that each predicate's extension is made up of a certain number of independently identified individuals—for most predicates are such that nobody could ever be in a position to have identified all the requisite individuals.

One way in which the requisite analysis of scientific linguistic practice might be developed is suggested by the discussion of the first two chapters. It emerged from that discussion that scientific predicates were applied according to two kinds of procedures. On the one hand were '*inferential* procedures' sustained by the acceptance of lawlike generalizations, which allowed the application of predicates to individuals, conditional on the applicability of other predicates to those individuals. Then there were '*observational* procedures', which govern applications of predicates in direct response to certain kinds of sensory stimulations. (It is worth pointing out that a distinction between two kinds of *procedures* for applying predicates is not equivalent to a distinction of the double language kind between two kinds of *predicates*. For, as observed at the end of Chapter 1, most, if not all, predicates will be involved in both kinds of procedures.)

It is clear enough how the applicability of predicates to 'new' individuals could be accounted for by reference to these two kinds of procedures. A predicate might be applied to a 'new' individual as a direct result of the reception of certain sensory information. Alternatively, an inferential procedure could lead to a predicate being applied to something on the grounds that it satisfies certain other predicates (though this latter kind of application will of course always be contingent on an observational procedure being used at some point).

Given this picture of scientific language, why should we not now consider a predicate's extension to comprise just those individuals which the established observational and inferential procedures of the language could result in its being applied to? This promises to give us a real hold on the notion of a predicate's extension, by reference to which we might be able to attempt a serious evaluation of the referential approach to problems of meaning variance.

But there are difficulties involved with this conception. For one thing, there is a problem analogous to that of the 'underdefinition' of 'theoretical' terms in the double language model. We take it, for instance, that any particular object in either (water-) soluble or it is not; that is, that the extensions of 'soluble' and 'not soluble' together exhaust the universe of objects. But there seem to be plenty of objects to which neither of these predicates could justifiably be applied, simply for lack of their ever being tested in any relevant way. Which, given the suggested account of what is required for an individual to fall in the extension of a predicate, implies that 'soluble' and 'not soluble' are not mutually exhaustive after all.

But this difficulty is not decisive against the suggested account of extensions. For there is always room to question whether all objects are indeed determinately soluble or not, independently of their being tested in some relevant way. Thus Dummett for instance suggests that in such cases the extension of 'soluble' need not be considered as always having been definite, but as somehow emerging as and when relevant tests are developed and applied. (Dummett's ideas will be discussed further at the end of this chapter.)

A far more pressing problem is the converse one, analogous to the earlier difficulties about '*over* definition'. Our suggested account of what fixes a predicate's extension clearly presupposes that the procedures in a given language will never lead to both a predicate and its negation being applied to a given particular. For, given that we cannot rest with a contradiction, if such a case were to arise our account would not specify whether the particular in question is the predicate's extension or not. However it is easy enough to see that such cases do arise. Consider any instance of what I shall henceforth call a scientific *anomaly*: a case where one set of observations and theoretical inferences leads to a singular assertion while another set of observations and inferences leads to the contrary assertion. The simplest case could be where observationally judged initial conditions together with accepted generalizations led to a predictive application of some predicate to some particular, while direct observation produces the negation of that prediction. But there will be other cases where both of the inconsistent statements involve the use of inferential procedures as well as observational ones.

The possibility of such anomalies effectively undermines the proposed analysis of a predicate's extension. The point can be illustrated most clearly by an example. Consider the question of how the results of the Michelson-Morley experiment were to be described in the language of the classical mechanics of the time. The experiment used a certain pair of rigid bodies ([x,y]) which were such that measuring them by rigid rods gave the same numerical results. Given the then accepted physics, this warranted the application of the predicate 'are congruent' to [x,y]. But it was also the case that light rays took the same time to traverse the two bodies when they had different motions. And this, again given the then accepted physics, warranted the application of 'are not congruent' to [x, y]. So: was [x,y] in the extension of 'are congruent' as used in the contemporary physics, or not? There is nothing in the analysis so far which suggests how this question might be answered. And the difficulty is clearly general, arising whenever one combination of procedures leads to an assertion Pa and another leads to $\sim Pa$.

It is worth emphasizing that the problem here is not simply a matter of vague predicates. What I am pointing to is not simply that the procedures for applying scientific predicates are sometimes too loosely formulated for it to be clear how they apply to certain cases. Rather it is that the application of the procedures can in itself be quite straightforward, yet still lead to the unwanted result that an individual satisfied both a predicate and its negation.

The difficulty I have raised here is directly relevant to the evaluation of the referential approach to the relativist problems deriving from the meaning variance thesis. For the postulated indeterminancy in the extensions of scientific predicates arises at just the point where the referential values of terms in different theories becomes an issue for scientific theory-choice. Consider again the example of the Michelson-Morley experiment. If [x,y] was in the extension of 'are congruent', as used in the language of the current physics, then '[x,y] are *not* congruent' would be false, and the correct response to the experiment would have been to reject this statement. This was what was done by special relativity, by virtue of its revising the classical assumption of the varying velocity of light in different reference frames. On the other hand, if [x,y] was not in the extension of 'are congruent' then '[x,y] *are* congruent' would have been the false assertion, and it would have been the Lorentz theory, which rejected this assertion by revising assumptions about the relation between rigid rod measurements and length, that constituted the right response to the experiment. But if there is nothing which makes one rather than the other of these analyses the right one, then it seems there is nothing which makes one rather than the other of the two resulting theories correct.

Again, the difficulty is quite general. It arises whenever the procedures for using a scientific predicate produce an anomaly. If it is not determinate in such cases which of the two anomalous applications of the predicate in question is false, then it seems that there is nothing to make it the case that one way of dealing with the anomaly is right and the other wrong, nothing to make it the case that it is right to revise the procedures leading to one of the anomalous applications rather than those leading to the other.

At this point I want to consider the obvious objection to my diagnosis of the situation. It could be argued that it is simply illicit to assume, as I have done, that all 'inferential' and 'observational' procedures contribute to the determination of predicates' extensions. For one thing, this seems to make any accepted generalization 'true by definition', by in effect laying down that the extensions of all the predicates it involves are related in just the way required to make it true. And my suggestions similarly make any observational procedure perfectly reliable, by stipulating that the extensions of the predicates it involves include all particulars to which it could lead that predicate to be applied. Surely I ought to leave room for the possibility that some accepted generalizations might be false, and that certain observational procedures might be unreliable. And is it not just this insistence that *all* generalizations and observational procedures are alike guarantees of their own reliability that led to the collapse of the proposed analysis of extensions?

The suggestion implicit here is that there is some canonical sub-set of observational and inferential procedures, and that only those particulars to which this set of procedures could justify a predicate's application should be deemed to fall within the extension of that predicate.

This strategy promises to resolve the difficulty raised by the possibility of scientific anomalies. For if there were some such privileged sub-set of rules, it might well be that whenever two anomalous applications of a predicate and its negation arise, only one of them could be justified by that sub-set of procedures alone. And this would resolve the question of whether the particular in question was in the predicate's extension or not. Moreover, it would rehabilitate the referential approach to the meaning variance thesis. For, if it is always a determinate matter which of two anomalous sentences is true, then it will always be objectively correct to deal with such anomalies by making those revisions which eliminate the false sentence, rather than ones which eliminate the true one.

An initial objection to the idea that the extensions of scientific predicates are fixed specifically by a privileged sub-set of procedures is that it is not clear why even this should guarantee that such predicates will have determinate extensions. For why should not the privileged sub-set of procedures on occasion produce an anomaly within themselves? That is, what is to rule out the possibility that these privileged procedures alone will lead to both a predicate and its negation being applicable to the same particular? And if this does happen, then of course we will still be faced with the original problem as to whether the particular is in the predicate's extension or not.

It might be felt that, if an anomaly did arise within the privileged sub-set of procedures, this would simply show that the privileged sub-set must have been misidentified. But why should there have to be some such 'anomaly-proof' sub-set of procedures? Traditional empiricism had a predicate applying to just those things that would produce the associated sensory impression on inspection. On this account there could of course be no two ways about whether a given particular satisfied a given predicate or not. But if we are to allow that a number of different procedures can be on a par in governing how a predicate works, then there is no automatic argument to exclude the possibility of its being indeterminate what a predicate's extension comprises. It seems likely that our unreflective tendency to take it for granted that a predicate's extensions are determinate is no more than an outmoded hangover from traditional empiricist and related theories of meaning.

Indeed, if the privileged sub-set strategy really were to guarantee the determinacy of predicates' extensions, then in the end there would have to be no more than one way in which the privileged procedures could lead to the application of each predicate. (Recall the discussion of operationalism in Chapter 1.) But we need not pursue this line of thought, for there remains a rather more fundamental objection to the privileged sub-set strategy. Namely, what is supposed to distinguish the privileged procedures (or rather, procedure) for a given predicate? At bottom what is being asked for here is something that will select out analytic generalizations, and analogously inviolate observational procedures, from the rest.

When we look at it like this it becomes clear that any arguments against the analytic-synthetic distinction will be equally applicable against the supposition that there are privileged procedures which alone fix the extensions of predicates. Thus we have been led round to the conclusion that the extensions of scientific predicates are no less indeterminate than their senses. Unless more can be said it seems that turning from the theory of sense to the theory of reference will be of no help at all with the meaning variance thesis and the attendant relativist problems. At the end of Chapter 2 we were in the position that (short of supposing an analytic-synthetic distinction) we could not get a tight enough hold on the notion of 'sense' to decide whether or when the senses of scientific terms did change with meaning change. Scheffler argued that this was of no consequence, since the objectivity of science required only that the referential values, not the senses, of scientific terms remained constant through theoretical change. But this in turn is of no consequence if it is as difficult (and for the same reasons) to understand what it is for the referential values of scientific terms to remain constant as it is for their senses.

3 ALTERNATIVES TO VERIFICATIONISM

At this point I wish to consider a general objection to my overall line of argument: namely, that I have so far taken a *verificationist* conception of meaning entirely for granted.

The logical positivists explicitly equated the meaning of a sentence with the method by which that sentence could be observationally verified. ('The meaning of a proposition is the method of its verification', Schlick [1936].) As we saw in Chapter 1, this equation of meaning with verification ran into difficulties, initially with 'theoretical' terms, and more generally in connection with the distinction between 'observational' and 'theoretical' terms itself. But I continued in one sense to take this equation for granted—the difficulties raised were precisely that it was unclear what did verify statements involving 'theoretical' terms, and indeed what verified even statements made observationally.

Again, in Chapter 2, I suggested that a theoretical context account of meaning was plausible precisely because the acceptance of theoretical postulates seemed to play a predominant part in fixing what counted as a verification of singular sentences made using the predicates involved. And when it seemed unlikely that all accepted generalizations could play this role, I wondered whether some sub-set of them did so. The reconsideration of observation and meaning at the end of Chapter 2 went along similar lines: the thought was that perhaps some observational procedures laid down verification conditions for statements, even if it were not true that all did so. The impasse in both cases was the apparent impossibility of substantiating the supposed distinction between inferential and observational procedures which played a part in fixing verification conditions, and those which did not.

Indeed it could with some justice be complained that in the present chapter too I have been working, albeit implicitly, with an essentially verificationist conception of meaning. My immediate concern has been with what it is for scientific predicates to have the extensions they do (in so far as they determinately do have any such), and, in line with Scheffler's suggestion, I have been trying to identify these independently of any explicit characterization of their senses. But there remains the point that on the standard Fregean conception it is the meanings (senses) of predicate expressions that gives them their extensions. And what I have been presupposing is that in so far as anything gives extensions to predicates it is the observational and inferential procedures for verifying assertions in which those predicates are applied to particulars. So once again I could reasonably be accused of simply taking it for granted that the meaning of a sentence is to be equated with the procedures for verifying it.

So at this point I shall consider some alternatives to a verificationist conception of meaning. Perhaps the way out of our present predicament with the meaning variance thesis and its attendant relativist difficulties will become clear once we remove our verificationist blinkers. In the next two sections I shall consider the view that the meaning of a sentence is given, not by the conditions which count as verifying it, but by the belief it expresses. And after that I shall turn to the idea that meanings are a matter of 'truth conditions' rather than verification conditions.

4 BELIEFS AND RADICAL INTERPRETATION

The idea that statements have meaning by virtue of *expressing beliefs* is a popular and long-standing view, and one that has had a recent revival. But what exactly is it for a sentence to *express* a belief? H. P. Grice has argued that the primary notion here is that of what a given speaker means on a given occasion by an utterance. A speaker so means something when he intends to produce a certain response in an observer by means of the observer recognizing that the speaker's intention is to do just that. Grice's analysis is designed to deal not just with assertions, but also with questions, commands, wishes, etc. These different kinds of utterance can be distinguished by the different responses the speaker intends to produce. It is specifically assertions—those utterances that raise the question of truth and falsity—that are of interest to us here. For assertions, Grice's analysis is that a speaker means something when he intends his observers to conclude that he (the speaker) has a certain belief by means of their recognizing that his intention in speaking is to get them to conclude that. (Grice [1957], [1968].)

What is meant by an utterance in Grice's primary sense can well vary from occasion to occasion for tokens of the same assertion-type. (The point here is not indexicality or ambiguity, which I shall continue to ignore, but the possibility of puns, irony, new metaphors, ignorance of conventional meaning, etc.) What we need to know is what it is for an assertion-*type* to have an established meaning in a given linguistic community. How is *this* supposed to be a matter of that assertion-type 'expressing' some belief? Recent work has suggested that this might be explicable in terms of a *convention* or *common expectation* in the relevant linguistic community connecting the assertion-type with the belief. There are a number of competing ideas about how exactly this should be spelt out. One is to say the convention obtaining if everybody knows that everybody knows that everybody knows that . . . that everybody will utter that expression only if they want their hearers to believe they have the relevant belief. However this does seem unnecessarily (indeed impossibly) complicated. An alternative suggestion is that it will do to have nobody doubting that nobody doubts. . . . And (for different reasons) it seems enough to have what they don't doubt being simply that everybody will utter that assertion only if they have the relevant belief. (Cf. Grice [1968]; Lewis [1969]; Evans and McDowell [1976]; Peacocke [1976].)

However we can leave to one side these and other such complexities as are occasioned by the worry that the notion of a convention or common expectation requires people to have implausibly sophisticated attitudes about each other. For the 'belief theory of meaning' often arouses a much more fundamental disquiet in philosophers of language: even if speakers do have the sophisticated attitudes required, is there not something inadmissibly circular about analysing the meanings of assertions in terms of the beliefs they express? The thought here is that the notion of belief is conceptually posterior to that of meaning, and so cannot play a part in explaining it.

There is indeed some substance to this accusation of circularity. But it is not, I think, fatal to the belief theory of meaning. What it will mean, however, is that the belief theory of meaning will be of no particular assistance in dealing with the meaning variance thesis.

The best way of bringing out the issues involved here is by considering the problem of radical interpretation. Let us consider a standard radical interpreter trying to understand the assertoric practice of some radically alien linguistic community. (Cf. Quine [1960], Ch. 2.) And let us help him initially to hypotheses about which utterance-types are assertion-types, and also about which tokens of such types are expressions of rational beliefs by sincere, literal, and competent speakers. (I shall say a speaker is *competent* if he knows what the normal understanding of his assertion is. A *literal* speaker is one who expects to be so normally understood in the particular case; that is, who is not being ironic, creating a metaphor, or otherwise playing with words. A speaker is *sincere* if he is not trying to mislead his hearers.) In the end, as we shall see, it will be possible for the radical interpreter to make a kind of check on his initial hypotheses and revise them accordingly.

It might now seem a simple matter to decode the meanings of the alien assertions—the meaning of an assertion-type will be that belief that is common to all those rational believers who utter tokens of it sincerely, literally, and competently.

But how is the radical interpreter to know who holds which beliefs when? Our normal way of working out what people believe is by noting what assertions they make and imputing beliefs accordingly. This route is obviously blocked if it is not yet known what their assertions mean.

An alternative strategy would be to try to work out their beliefs solely from their non-linguistic behaviour. That is, the interpreter could suppose that as functioning members of *homo sapiens* the aliens perform those actions which their beliefs indicate as the ones most likely to optimize the satisfaction of their desires; and he could then try and use this supposition to work backwards from what they do to what they must believe (and desire). But this will not do either. In principle there will obviously always be an infinity of logically possible combinations of beliefs and desires which might account for any given set of actions. If the aliens' non-verbal behaviour were really the only thing he had to go on our interpreter would be at an entire loss as to how to proceed.

Perhaps the interpreter should try getting at what the aliens believe via the causes of their beliefs, rather than via their effects. That is, he could consider what their various individual circumstances, their various past experiences, must have led them to believe, instead of trying to work out what beliefs must have led them to act and speak as they do. For instance, to take a fairly trivial example, he might feel entitled to conclude that an alien looking at drops of water falling from the sky will believe *it is raining*.

But do we know what beliefs the aliens will come to give certain peripheral sensory inputs? Perhaps their concepts are different from ours, perhaps they make different things of their peripheral stimuli. Our aliens might for instance have various exclusive categories into which different examples of our rain fit—say one for mist and light rain up to a certain point, and another for heavier rain, snow, and hail. So it might be that on looking at water falling from the sky they do not after all form any mental judgement which can be identified with our *it is raining*. Other less trivial examples of alien cultures using different concepts from ours suggest themselves 'bewitched', 'homunculus', 'dephlogisticated'. (The possibility I am raising here is of course little different, *modulo* the belief theory of meaning, from the possibility of meanings varying from scientific theory to scientific theory. But let us continue with the quasi-anthropological model for the moment.)

The possibility of conceptual variation across cultures is a somewhat controversial one, not least because of the just-explained difficulty it raises for the enterprise of radical interpretation. But, abstracting for the moment from any philosophical difficulties that may follow, it is very natural to suppose that the different psycho-developmental trainings to which people in different cultures are subject lead to their having different mental make-ups. In particular it is certainly initially attractive to suppose that different linguistic trainings will produce different conceptual abilities. On this model a linguistic training does not just associate words with preform types of beliefs, but also develops the ability to form certain beliefs: being conditioned to use sentences in response to certain sensory promptings and according to certain inferential principles can itself instil in the individual the concepts appropriate to that sentential usage.

Before pursuing the question of cross-cultural conceptual variation it will be helpful to digress briefly and comment on the terminology of *concepts* which I have slipped into in the last couple of paragraphs. I have been speaking of a disposition to form beliefs with a certain content as a matter of having certain 'concepts'. The idea here is that concepts are somehow the building blocks out of which beliefs with given contents are constructed. If we allow, as I think we must, that any individual is capable of adopting any of an indefinite number of beliefs, then we must allow that he has a structured and generative mental capacity for forming beliefs. And so it is appropriate to think of the capacity to form beliefs as involving a grasp of the concepts out of which those beliefs are constructed.

It is worth emphasizing that concepts in the sense intended here are not sensory images of any kind. The arguments for the judgemental account of sense experience are relevant here. There is obviously something imagistic about our sensory experience, especially our visual experience. But the concepts in terms of which we respond to our sensory experiences by forming beliefs, and indeed in terms of which our sensory systems produce beliefs *in* our experiences, always transcend the imagistic contents of those experiences. To judge, visually or otherwise, that something is a rectangular space, is to judge something more than that it produces a certain visual image—for, as instanced by the Ames room, there is the possibility that just that image might be produced by a non-rectangular space. Nor will we find imagistic concepts even at the level of how things appear. To hold that something *appears* so-and-so is not to form a belief different from one to the effect that it *is* so-and-so—it is simply to form the same belief in a different 'place', namely in one of one's sensory systems instead of at some higher mental level.

Let us now return to the main line of argument. As we saw, the possibility of cross-cultural conceptual variation raises a serious problem for any attempt at radical interpretation. In itself this is scarcely a reason for dismissing this possibility out of hand. But if we do admit the possibility, how then is the radical interpreter to proceed?

It might be thought that he should avail himself of some 'principle of charity' at this point: will he not do best to translate so as to maximize the ascription of (what we hold as) true beliefs to the various alien asserters? I shall have more to say about principles of charity in

Chapter 6. But in the present context the suggested strategy is clearly an *ad hoc* hotchpotch. The idea that the aliens' beliefs should in some sense correspond to ours only makes sense if we assume that at bottom they share our concepts. But then the admission that the fit is not going to be perfect seems to allow that perhaps they have different concepts after all. (Remember I am assuming that—*modulo* their concepts—all the aliens' beliefs are rational, and so am bypassing the more usual reason for thinking some of the aliens' beliefs will fail to correspond to what we hold as true.) Suppose that the interpreter's aliens actually did have different concepts from his. If he is forced to select meanings for their assertions from his possible beliefs, then he will inevitably find himself unable to explain their saying and believing at least some of what he takes them to be saying and believing, even though in itself their (variant) cognitive practice is quite coherent and understandable. Surely it would be preferable—and indeed more charitable—if it were possible to proceed without a principle of charity of the kind suggested.

I think this is indeed possible. But the discussion so far should have made it clear that it will not be possible to identify the aliens' beliefs independently of interpreting their assertions. The interpreter cannot first fix their beliefs via an examination of their individual circumstances and then see what belief is common to all those who utter a certain form of words. But there is an alternative strategy.

The suggestion made a few paragraphs ago was that certain concepts might be instilled in individuals precisely by their being trained to use given sentences in response to certain sensory promptings and according to certain inferential principles. This raised the possibility that different linguistic communities might have different sets of concepts. But if alien concepts vary for this reason (or indeed for any other reason) then will not this variation be reflected in the structure of the alien assertoric practice? And so why should we not use the structure of that practice itself to show us what contents they attach to their beliefs, what concepts they have? In a sense what we have to do is take *simultaneous* account of their various individual situations and their various assertoric outputs, and use this to attach such contents to their assertions and beliefs as will tally both with what each turns out to say and with what each must believe. So for instance our interpreter might notice that there is an assertion-type which is uttered only by those who have seen someone having a certain kind of convulsive fit or by those who believe that that person was the target of a certain spell.

This will then tell him that the assertion in question expresses a concept of a certain individual state, namely one which is observably evidenced by fits of the relevant kind and which can be caused by a certain kind of spell. (I am, it is true, here supposing that our interpreter will already know certain of the aliens' other beliefs, such as those about who is the target of what spells. But this shows only that his interpretation will to some extent have to proceed holistically.) The idea is thus that we correlate the sensory evidence and inferential premises of individual aliens with the assertions they variously make; and then we fix the belief which an assertion expresses as that belief which can be warranted by grounds possessed by all those who make it.

Once the radical interpreter has completed his task he will have found out what concepts the aliens work with (as well as finding which words express them). And then, of course, he would be able to judge what belief a particular alien would come to in given circumstances, independently of whether the alien expresses that belief in an assertion. But such judgements cannot be used as an independent means to working out the meanings of alien assertions in the way desired earlier. For the interpreter is in a position to make such direct judgements of what individual aliens must believe only after he has identified their concepts, and this identification is itself part and parcel of interpreting the meanings of their assertions.

Does the interpretational strategy I have outlined really get round the possibility of cross-cultural conceptual variation? In particular, am I not implicitly assuming that the interpreter already shares the alien concepts, in supposing that he will succeed in attaching unitary contents to their structurally strange uses of terms? Or, if not that, am I not at least supposing that he already has concepts out of which he can construct such contents? For instance, how can I be sure he will be able to identify which kind of convulsion justifies applying the relevant epithet, unless he already has the concept of that kind of fit? However, all that is required here is that the interpreter be able to acquire a grasp of the requisite concepts, not that he have them beforehand. If acquiring a first language instils new concepts in us, then why should a further language not do the same? If it is possible to learn a first language, then it should be possible to learn further languages underlain by different concepts. Of course there will be obvious difficulties about translating from one language into a conceptually divergent one. But there is no particular reason to suppose that understanding a new language requires being able to translate it into a language one already speaks. (I am not of course saying that different languages *have* to be underlain by different concepts: if the structure of experiential associations and inferential principles for the words in two languages is the same, as it no doubt almost entirely is with, say, modern English and modern German, then the concepts behind those words will be the same, and there will be no particular barrier to straightforward translation.)

No doubt there are ultimate constraints on the ability of human interpreters to develop coherent interpretations for sentential usages. Indeed it is essential for the interpretational theory I have outlined that there be some such constraints: if there were none at all then there would be nothing to stop an interpreter constructing concepts to make rational sense of any repeated sounds produced, however arbitrarily, by an organism. But we can allow that there are some such constraints without concluding that an interpreter's concept-developing abilities will always be exhausted by the demands of his mother tongue. (In so far as the constraints on humans' concept-developing abilities are universal, they will in any case be no hindrance to the radical interpretation of alien communities—for any constraints to which the interpreter is subject would have applied also to the interpretees when they were originally acquiring their language as children.)

There is however one difficulty about the acquisition of new concepts which needs facing up to. I have suggested that the possession of certain concepts involves being disposed to form certain beliefs when apprised of certain grounds. Does this mean that somebody who learns to understand some traditional society is going to end up actually believing the same things as they do in response to given sensory and other information? There is no great problem about learning a new way of categorizing meteorological precipitations, say, or going along with the Eskimos and the Bedouin in their complex respective classifications of snow and sand. But what happens when the aliens have concepts involving witches and spirit possession and so forth? Is the poor anthropologist to be forced, in acquiring those concepts, to end up actually believing that individuals who evidence certain behaviour are bewitched? Does the historian of science have to believe that mercuric oxide is dephlogisticated mercury? The trouble is that beliefs such as these, unlike Eskimo snow beliefs and Bedouin sand ones, carry an ontological and theoretical weight which is inconsonant with our own world view. Surely coming to know that other people hold such beliefs cannot require that we should give up what we have independent reason to accept?

Now one possibility here is to distinguish between grasping what certain alien concepts are and actually having those concepts. The distinction would be between knowing that their mental make-up is such as will account for their using words according to certain identifiable observational and inferential uniformities, and actually oneself having a mental make-up which leads one to use words naturally in that way. Perhaps, so to speak, we could suppose that the anthropologist acquires 'concepts of their having certain concepts', without himself going on to acquire the latter concepts as well: he could know what observational and inferential procedures manifested their conceptual system, without himself having 'internalized' those procedures and therewith that system.

The distinction being suggested here is clearly not a sharp one—how facile does one have to get at using words according to certain principles before one moves from just knowing what those principles are to having 'internalized' them? Still, it clearly has some substance, and it is plausible to suppose that what instils concepts is linguistic practice, not mere knowledge of linguistic principles. On the other hand, this line of thought is not going to solve the problem at hand. The anthropologist does not, indeed surely could not, set himself to comprehend his aliens' language purely 'externally', simply giving a detailed structural description of the way they use sounds. Even if historians of science do something along those lines, the anthropologist proceeds to learn the language like a native; learns *how* they use words, not just what the uniformities governing their usages are. But then, if such first hand linguistic trainings instil concepts, and concepts are dispositions to form beliefs, it ought indeed to follow that the anthropologist ends up forming beliefs on the same pattern as do the people he is studying. Fortunately for my story, there is some indication that this phenomenon does actually occur—Western anthropologists doing fieldwork in traditional societies do often report that they experience a tendency actually to adopt the thought patterns of their hosts. This of course results in a certain schizophrenia, given the incompatibility of the anthropologist's new 'beliefs' with the intellectual baggage he has brought from his home culture and language. And given his commitment to his previous habits and standards of thought the anthropologist will normally resolve the conflict (no doubt quite rightly) by rejecting his new-found 'beliefs'. But how such conflicts are resolved is not crucial to the point at issue—it seems a satisfactory enough corroboration of my general line of argument on radical interpretation that such conflicts do indeed arise in the first place.

A related point. It is often complained, against particular putative examples of conceptual variation, that the very presentation of the examples belies the thesis at issue. For instance, 'witch' (and 'dephlogisticated') are expressions of twentieth-century English, and surely as speakers of that language we understand them perfectly well. But if that understanding consists of a grasp of the concepts expressed, then why ever should it be supposed that we lack certain concepts possessed by, say, the Azande (or by eighteenth-century chemists)? However, the important point here is that we do not use the words in question to describe reality, except for the specific purpose of describing the content of other peoples' thoughts and assertions. *We* don't ever conclude that someone is literally a witch; only that other people have sometimes concluded that. In a sense words like 'witch' (and 'dephlogisticated') serve the quite specific purpose of extending contemporary English to make it adequate to represent the thought of historically or geographically distant communities. So what we understand by 'witch', as is shown by what beliefs we use this word to express, is a 'concept of others having a certain concept': this is not the same as, though it is obviously related to, the concept possessed by people who actually believe in witches. What would show that we had the same concept as the Azande would be our following them in actually using the relevant word to describe individuals who evidence certain characteristics. But of course we do not.

Perhaps a few further remarks will help to clarify my position here. It is after all fairly natural to suppose that we do mean pretty much the same by 'witch' as the Azande mean by their expression 'ira mangu'; that we have the same concept of *witch* as they do. Surely, like them, we have a certain idea of what it would be for someone to be a witch, what characteristics such a person would have. Why should the difference between them and us be anything more than that they take it that certain people actually do have those characteristics, whereas we know better?

On this view of things, that the Azande have different beliefs from us is not due to their having different concepts, but simply a matter of their diverging from us on whether there is ever adequate evidence for the instantiation of those concepts. My position contrasts with this natural view of things. I do not simply want to say that what is different about the Azande is that they habitually leap to conclusions we know better about: on the contrary I would hold that, given the (different) concepts they think with, such judgments as 'so-and-so is a witch' are characteristically no less warranted than any singular judgments we or any other peoples make.

The full rationale for preferring this latter kind of account will have to wait until the latter part of Chapter 6. But lest it seem overly implausible to start with, consider briefly the way the Azande actually do use the term 'ira mangu'. Amongst other things, they hold that anybody with a certain 'witchcraft substance' ('mangu') in their abdomen is a witch, and on occasion they discover this substance after a post mortem on someone. They also of course hold that witches can effectively wish misfortunes on their enemies.

Now if we concentrate on this last facet of their notion, then it will seem that they are mistaken whenever they infer witchhood from post mortems—for presumably people who contain the relevant substance are not in fact able to visit misfortune on their enemies at will. But suppose we concentrate instead on the inflated gall bladders (assuming, after Evans-Pritchard, that that is what 'mangu' is) as specifying what it is to be an 'ira mangu' (as indeed the etymology suggests). The Azande will then characteristically be quite right when they conclude someone to have been a witch, for there are people who turn out to have inflated gall bladders. The Zande mistakes would then come out in their further supposing that those witches can visit misfortunes on their enemies. But of course there are not really two possibilities here, not really two different ways in which the Azande might be erring—there are just two artificially different ways of describing the same situation. So it is after all no simple matter to diagnose the Zande error, to pinpoint exactly where it is that they are unwarrantedly jumping to conclusions. Clearly there is something about Zande linguistic and intellectual practice that gives us reason not to adopt it ourselves. But it will prove easier to explain exactly what this is if we take it to be a matter of their thinking normally with inferior concepts, rather than, as the natural view has it, a matter of their thinking inadequately with concepts which we share.

5 WHY BRING IN BELIEFS?

In the course of this long discussion of radical interpretation it should have become clear why a belief theory of meaning might be accused of circularity. I have suggested that having certain concepts is in effect for there to be certain sentences such that one is disposed to accept them if apprised of certain grounds; to actually form a given belief is correspondingly then to accept some sentence to which one is so disposed. (Accepting a sentence is itself another disposition—namely, one to assent to the sentence when prompted.)

Now, if all this is indeed right, then we can only get at the identity of a belief, at what it is for the belief to be *that* belief, via what grounds are accredited as warranting such sentences as express it. So surely we only understand the notion of *belief* by virtue of a prior understanding of what it is for a sentence to have a certain established assertoric use in the language. And so would we not have cut a long story much shorter by simply sticking with the old Viennese verificationism which equated the meaning of an assertion directly with the evidential grounds which warrant its acceptance? What does talk of 'concepts' and 'beliefs' add to our understanding of how language works?

A first response to this accusation might be to say that even if judgments about beliefs and concepts are epistemologically dependent on the data of language use, it does not follow that more judgments are either *conceptually* or *explanatorily* so dependent. That is, even if we can only find out about people's beliefs and concepts via the patterns displayed in their use of language, we can still conceive of their having those beliefs and concepts as something distinct from their linguistic behaviour, and indeed as something that might explain that behaviour. (Cf. Evans and McDowell [1976], p. xvi.)

But, even so, it is still not at all clear what is really to be gained by bringing in beliefs and concepts. Allowing that a theory of meaning should, amongst other things, facilitate explanations of verbal behaviour, do we not do better simply to explain verbal utterances or assents (S) straight off by reference to current prompting (P), the prior acquaintance with accredited evidential grounds (G), and the still earlier linguistic/cognitive training (T) (together with the 'behaviourist' generalizations of the form (T $\rightarrow$ (G $\rightarrow$ (P $\rightarrow$ S))) which we implicitly assume in any case)?

The real reason for favouring a belief theory of meaning becomes apparent only when we drop the pretence that people always express rational beliefs and do so by speaking sincerely, literally, and competently. For once we do look beyond the limits of the radical interpreter's initial hypotheses we have to acknowledge that there are plenty of cases of linguistic behaviour which cannot be accounted for as authorized responses to accredited evidential grounds and current prompting.

Why should this matter? Cannot the 'behaviourist' account for the existence of linguistic deviations easily enough, as somehow due to the absence of, or interference with, the linguistic/cognitive training which is required for grounds and prompting to ensure the correct linguistic response? However, the crucial issue here is not so much just that cases of linguistic deviation exist, but rather that we want to distinguish amongst various *different* ways in which a speaker might deviate.

Someone's beliefs might be quite in order and yet his words might fail to express them—this would be a case of insincerity or unliteralness. Alternatively, his words might express his actual beliefs all right, even though those beliefs resulted not from his contact with accredited evidence, but instead from wishful thinking, or from the need to rationalize certain of his actions, etc.—this would be a case of beliefs arrived at irrationally. These two kinds of case are clearly quite different from each other, and are moreover both different from what is the yet further case of an incompetent speaker who deviates because his initial linguistic/cognitive training failed in the first place to develop in him a grasp of the right concepts and the words to express them. Yet on the 'behaviourist' theory these three kinds of deviation all collapse into one. The real strength of a belief theory of meaning is that it allows us to discriminate between these importantly different kinds of linguistic deviation, and moreover does so in a manner which allows for an appropriate continuity between explanations of deviant and non-deviant utterances. (Cf. the argument for a 'sense experience' account of observation reports in Chapter 1, p. 32.)

To digress briefly, it might seem odd that I have just one category of 'incompetence', that I am talking undifferentiatedly of such incompetence as a lack of 'linguistic/cognitive' skills. Surely there ought to be some room for the possibility that an individual has certain concepts but simply lacks the words to express them?

This possibility would indeed be ruled out if it were true that given conceptual abilities always resulted from and were hence reflected in a corresponding linguistic practice. But if we are to allow, as I think we must, that people (and other organisms) can sometimes have concepts they cannot express, then we should recognize that this generalization cannot be strictly accurate. One possible exception would be where someone forgets which words express a certain concept. That is, he might keep a certain concept while losing the linguistic habit which first gave it to him. More fundamentally, there is the further possibility that certain concepts are innate (simple geometrical concepts, say), or are such that they can be produced by environmental influences other than linguistic trainings.

So there is no great difficulty about conceptual power exceeding linguistic display; nor, correspondingly, about distinguishing, amongst cases where individuals or groups have different 'linguistic/conceptual' abilities, those where they diverge in both word and thought, and those where the deviation is merely in respect of linguistic resources and does not reflect any conceptual differences. There are of course problems about how to *justify* the attribution of unexpressed concepts to people. Presumably such attributions will depend on specific theories about how such concepts can get instilled in people and how they might then show themselves. But in any case I shall ignore unexpressed concepts henceforth, and continue to speak undifferentiatedly of 'linguistic/cognitive' abilities. For our primary concern is with linguistic meaning, which we are now supposing to be a relationship between words and such concepts as *are* expressed by words: given this concern it matters little whether variations in such relationships are due to the absence of a word to express a concept or the absence of the concept itself. (Note also that nothing in the earlier analysis of radical interpretation presupposed that every conceptual ability would depend on an appropriate linguistic training. The possibility of some such dependencies was, it is true, used to show that in posing the problem of radical interpretation we need to allow for the possibility of conceptual variation. But the solution to the problem of radical interpretation assumed nothing in particular about where people's concepts come from in the first place, but only that the structure of their linguistic practice does reflect whatever concepts they actually do have and express.)

To return to the main line of argument, note now that the reasons which force the belief theory of meaning on us at the same time show that my earlier remarks about beliefs and concepts and about radical interpretation were over-simplified. Having a belief cannot after all be a matter of a sure-fire disposition to assent to some sentence when prompted; nor can having certain concepts be a matter of sure-fire dispositions to form beliefs when apprised of certain grounds. For one can have a belief and yet utter words which *modulo* standard usage belie it (insincerity or unliteralness); and one can have a concept and yet form beliefs involving it which are unwarranted (irrationality). Both these possibilities clearly raise difficulties for the straight-forward interpretational strategy of reconstructing the concepts behind an alien community's words by simply correlating which sentences they utter with the grounds they are variously apprised of. And then there is the further difficulty that through inadequate linguistic/cognitive socialization individual aliens might fail to be competent members of the alien linguistic community. However, these difficulties are not insuperable. What we should now recognize is that we need to allow for perturbations in the correlation of sentences uttered with grounds apprised of corresponding to cases of insincerity and unliteralness, to cases of irrationality, and to cases of incompetence. Of course we need some independent check here, to make sure the deviations from the postulated correlation really are as diagnosed, and not just artefacts of an initial misconstrual of the content of certain of the alien assertions. But it is clear enough how we might satisfy this demand. If we want to attribute an anomalous assertion to insincerity we need independent evidence of a desire to mislead; and if we want to attribute it to unliteralness we need independent evidence that the speaker did not expect to be under-stood normally. If we want to claim that the speaker had an irrational belief we need to substantiate some alternative explanation of his arriving at it despite the absence of good grounds. And if we want to say that a speaker lacks a proper understanding of the words he uses we need to show that there is something about him or his history which has prevented him from acquiring the same linguistic/conceptual abilities as his social colleagues. If we can find such independent substantiations whenever our interpretation implies that a particular utterance is deviant, then the interpretation can stand. If not, then of course the interpretation itself needs revising.

In effect this now shows how the radical interpreter can eventually have some check on his initial hypotheses. The radical interpreter starts with certain assumptions about who are rational, sincere, literal, and competent speakers. In accord with these initial assump-tions he comes up, if he can, with an interpretation of their assertions. This interpretation (and therewith the initial hypotheses it stems from) can then be tested by seeing whether it allows the supposedly deviant cases of assertion to be explained. This works as a test precisely because the explanation of deviant assertions will presuppose the imputed standard interpretation of the assertion-types involved. (Why should he have been insincere about *that*? Why should he have expected his hearers to have understood *that* ironically/metaphorically . . . in this context? What could have led him to believe *that* irrationally? What could have stopped him grasping that *that* is what is expressed by those words?)

It might be objected at this point that if we can have independent evidence for such attributions of deviance, as I am supposing, then the way must still be open to an analysis of linguistic behaviour as simply being uniform responses to external stimuli, albeit that the uniformities have now turned out to be much more complicated than hitherto supposed. The original reason for preferring a belief theory of meaning to a behaviourist one was that the latter apparently could not distinguish, say, insincerity from irrationality—it seemed to make both simply like deviations from established linguistic habit. But if we can substantiate attributions of insincerity, irrationality, and so on, then presumably we will do so by reference to certain antecedent occurrences of an external 'behavioural' kind (prior conditioning, previous behaviour, recent stimuli, etc.). And if these occurrences explain the linguistic acts in question, then presumably there is some (admittedly elaborate) system of generalizations framed in purely behavioural terms which will make out the differences between the different kinds of assertoric deviation.

I am not particularly concerned to take issue with this objection. There are various general reasons, in particular the demise of an authoritative observation language, for being suspicious of the kind of 'externalist' reduction being mooted here. On the other hand, special considerations arguably arise in connection with the relationship of mental events to physiological brain states which make a reduction more plausible in the special case of a mentalist theory of human action. But, in any case, my primary aim in this section has been to show that a theory of meaning must somehow be adequate to the distinctions we are normally able to make by taking assertions to express beliefs. This much at least has been shown, even if there remains the possibility that there might be alternative formulations also adequate to these distinctions.

So, to sum up the arguments of this section, a belief theory of meaning (or something very like it) is required if we are to explicate the differences between the various kinds of assertoric deviance. And the complexity that such deviance adds to the project of radical interpretation does not prevent the execution of that project, but indeed allows an independent check on the hypotheses which are needed to get it started in the first place.

Unfortunately, notwithstanding any independent interest they may have had, the arguments of this section and the last make it clear that the belief theory of meaning will not after all help with the meaning variance thesis and associated problems. For even if we take the meanings of words to be given by the beliefs or concepts they express, we still need to ask whether or not terms from different scientific theories in a given field do characteristically share meanings. Or, if not that, at least whether or not they characteristically refer to the same objective existents in reality. And as soon as we ask these questions from within a belief theory of meaning all the old problems arise once again.

In essence this is because the identity of the belief expressed by a given assertion is provided by the structure of established observational and inferential uniformities governing the use of that assertion. Even if people sometimes deviate from such uniformities, these are essentially derivative cases. People do indeed express beliefs they ought not to—either because they are not entitled to those beliefs, or because they do not have them—but what those beliefs are is still given by the authorized grounds for using the assertions that express them. Now—exactly which observational and inferential uniformities give the content of the belief expressed by some assertion? Every accepted lawlike generalization in a theory will sustain an inferential procedure for using the terms it contains. Presumably they cannot all contribute to the identities of the concepts those words express (to the meanings of those words), for then even the slightest change in theory will produce a wholesale change in meanings, and it will once again seem impossible for evidence to compel a change in theory. On the other hand there is as before no satisfactory way of dividing out those generalizations that do fix meanings from those which do not. And similarly with observational uniformities. Surely they cannot all play a part in showing what concept a word expresses, for it seems quite possible to discover that a certain observational procedure is not after all an acceptable basis for using a word (for judging something to satisfy the concept that word expresses). Yet, again, we lack any admissible means for discriminating those privileged observational uniformities that do manifest the identities of concepts.

Analogous problems arise if we switch to asking about the referents of scientific terms. If the meanings of words are given by the beliefs or concepts they express, then the referential values of those words will be those things which those beliefs and concepts are about. But if what those beliefs and concepts are is given in turn by the evidential uniformities to which the use of those words conforms, then we are effectively back with the same problems we faced in Section 2 of this chapter. If we take all the evidential uniformities in question then it seems doubtful that scientific terms (equivalently, the concepts those terms express) do have determinate referential values: some of those evidential procedures will indicate certain referential values, others will indicate different ones. And given that there is no privileged sub-set of such procedures there will be no apparent way of restoring determinate references to scientific terms.

('The views on the relation between language and thought which I have developed in this section and the last are not dissimilar to those put forward by Gilbert Harman in his [1973] and [1975]. There is, it is true, the apparent difference that Harman takes the identity of 'thoughts' to be given by the way they respond to perceptual evidence, by the role they play in inferential reasoning, *and* by the manner in which they give rise to action. But this difference is not deep. Harman's 'thoughts' include desires and other propositional attitudes, as well as beliefs. Because of my concern with assertion I have restricted my attention to beliefs alone. Actions emerge not from beliefs alone but from the interaction between desires (for certain ends) and beliefs (as to what means will achieve those ends). If we had wished to explicate what it was to (express a) desire that so-and-so, as well as what it is to (express that one) believe(s) it, then we would indeed have had to attend to the relation between propositional attitudes and actions. But given that our concern has been with beliefs alone this would have been redundant.)

6 TRUTH CONDITIONS

Perhaps it is not so very surprising that our foray into the belief theory of meaning has got us no further with the meaning variance thesis. For does not the belief theory as I have outlined it simply transpose verificationism into the mental realm? A belief theory of meaning is scarcely much of an alternative to a verificationist one, if the identity of beliefs is given by the evidential grounds for sentences expressing them—that is, by the verification conditions of those sentences.

So let us now look at another kind of alternative to a verificationist theory of meaning: namely the view which takes the content of sentences to be given not by their verification conditions but by their *truth conditions*. The idea involved here can best be brought out by considering once more dispositional terms like 'soluble' (or, to take examples made familiar in this context by Dummett, 'brave' [1958], 'good at learning languages' [1976]). At various points earlier in the book I have suggested that since many physical objects are never put to the relevant test, there are difficulties about construing 'soluble' as standing for a property which all such objects determinately either have or lack. In effect the argument was that the verification conditions for ascriptions of 'solubility' and 'insolubility' are not as such as to give us any warrant for supposing that every object satisfies either the one or the other of these terms. But such arguments can be made to cut a different way. Why not reason conversely that, since we do understand 'solubility' as a property that everything either has or lacks, the content of this term cannot be given just by the conditions which enable us in practice to verify whether something satisfies it or not? The idea here is that our understanding of 'soluble' must go beyond knowing how to *find out* whether something satisfies this term, to a grasp of what it is for something to be 'soluble' whether or not we are in a position to ascertain this in practice—that is, to a grasp of the truth conditions for saying something is 'soluble'. This argument, that meanings involve a grasp of truth conditions rather than verification conditions, can allow that for some assertions there is no substantial such contrast to be drawn. For assertions reporting, say, the colours of physical objects, it is arguable that we are always in a position to recognize effectively that such an assertion is true if it is. So the verification conditions and the truth conditions for such assertions coincide. But this equivalence breaks down as soon as we move to those assertions, such as those about something being 'soluble', which seem to have 'verification-transcendent' truth conditions—those assertions for which we are inclined to allow that there is no guarantee that we can recognize them to obtain whenever they do. It is specifically in these latter cases that the truth conditions theorist will insist that a mere specification of verification conditions falls short of an explanation of meaning.

The idea that meaning is given by truth conditions promises a new perspective on the meaning variance thesis. Since different theories will (by definition) differ on some general assumptions, they will inevitably indicate some different ways of finding out about the entities their expressions putatively designate. And so we have naturally enough found it difficult, given a verificationist approach to meaning, to explain why all theory changes do not amount to meaning changes. But if what gives assertions their meanings is rather an indication of their truth conditions, then there seems no immediate reason why meanings should always vary with theory. For the truth condition of an assertion can be a state of affairs which can obtain independently of the presence of any of the signs by which we hope to recognize it. So there seems room for us to change our minds (theories) about what those overt signs are, and yet still be indicating the same truth conditions—that is, there seems room for the meaning of an assertion to remain constant even while its verification conditions vary. For instance, if what it is to be 'soluble' is a matter of some underlying state behind any recognizable manifestations, then perhaps we will be able to change our minds about appropriate tests for 'solubility' without changing our minds about what 'solubility' is.

There are independent reasons for casting truth conditions in the central role in a theory of meaning. The meanings of sentences must depend systematically on the meanings of their components. The programme pursued by Donald Davidson and his followers promises to show, for all the iterative operators in a language, how the truth conditions of the complex sentences resulting from the application of such operators depend on the truth conditions of simpler sentences. (See especially Davidson [1967].) There are of course difficulties involved in carrying out the Davidsonian programme, in finding the requisite analyses of complex sentences. But the analogous programme, of showing how the verification conditions of complex sentences depend on the verification conditions of simpler ones, seems to face far greater (though not necessarily insuperable) difficulties. (See in particular Brandom [1976]; Wright [1976].)

On the other side, however, is an argument of Dummett's for preferring a theory of meaning based on verification conditions. (See in particular his [1976], Sec. II.) Dummett's complaint is in the first instance that it is fundamentally obscure how the association of an assertion with a 'verification-transcendent' truth condition, which by definition can obtain undetectably, can in itself do anything to illuminate the way language is used. If the truth conditions in question are ones which on occasion people are unable to decide the presence of, then what possible relevance can the association of assertions with such conditions have to the way people actually use those assertions?

If we construe truth conditions as possible states of affairs in the external world then this objection seems conclusive. An association of an assertion with an undetectable such state of affairs (if indeed such an association makes any sense) could not play any part in explaining actual linguistic practice. But suppose we see truth conditions instead (as I shall henceforth) as having a role within a belief theory of meaning—as specifying not certain possible objective circumstances associated with an assertion but rather the content of the belief that the assertion expresses. The switch from verification to truth conditions will then be a modification of the belief theory of meaning outlined above, not an alternative to it. And the argument for the switch will then be that we seem to understand verification-transcendent assertions as expressing beliefs the contents of which cannot be adequately explained just by specifying how we verify them—to grasp such beliefs requires recognizing that things might be as they make them out to be even when we cannot effectively verify that this is so.

On this understanding of the truth conditions approach Dummett's objection loses some of its force. For then ascribing truth conditions to an assertion would seem to relate straightforwardly enough to people's use of that assertion—people will utter or assent to that assertion just in case they wish to express or concur with the belief whose identity is given by those truth conditions. That they might lack an effective ability to detect the assertion's (belief's) truth conditions would no longer be of immediate relevance—what matters is that they have the belief, not that they can establish it.

But are things that simple for the truth conditions approach? Those suspicious about verification-transcendent truth conditions could at this point query whether there really is any reason to think our use of language is responsive to beliefs about possibly undetectable states of affairs: can we not explain assertoric language perfectly well without bringing in any such beliefs? Consider a community who are just like us in the way they make assertions in response to recognizable verification conditions, but who lack any further concepts of unrecognizable truth conditions. In what way does their use of language differ from ours? Presumably we actually make assertions only when we have established them, even if their content points to something more than the mere availability of recognizable grounds for doing so. So our linguistic practice would seem to be just like theirs. But then what part need any putative grasp of verification-transcendent truth conditions play in the explanation of our language? We will explain their linguistic practice without bringing in any such concepts, for by hypothesis they have none. But then why not ours too? For as just argued our linguistic practice is identical with theirs. Indeed this thought experiment suggests not only that concepts of verification-transcendent truth conditions should play no part in the theory of meaning, but moreover that it must be a mistake to think that anybody has such concepts. For what possible warrant could there be for attributing them to anybody in the first place? If their use of language cannot show they have them, what can? If this line of thought is right then it would follow that we are ourselves guilty of misunderstanding certain of our own concepts when we credit ourselves with notions of something being 'soluble', or someone being 'brave', without showing it. (Cf. Wright, [1976], pp. 231–4.)

However there is room at this point for an advocate of the truth conditions approach to dispute whether there is indeed nothing in our linguistic practice to show we attach verification-transcendent truth conditions to certain assertions. He can return to the kind of reasoning with which the idea of truth conditions as an alternative to verification conditions was first introduced. We are inclined to accept that everything is either such that it is 'soluble' ('brave', 'good at learning languages'), or it is not. That is (in effect), we are inclined to accept the law of the excluded middle, p v ~p, even for sentences such as '*a* is soluble or it is not the case that *a* is soluble'. Does this not itself show that we attach a verification-transcendent content to 'soluble'? For it is certainly not the case that everything can either be shown to satisfy 'soluble' or shown not to do so.

The suggestion here is that a linguistic community's acceptance of certain logical principles can itself provide the requisite evidence for attributing a grasp of verification-transcendent concepts to them: and in particular that their accepting the classical law of the excluded middle will do so. This line of argument can be explicated in terms familiar from Dummett's writings. (See in particular his [1974b].) Logical principles are to be accounted for by reference to the semantics of the logical constants such as v and ~. But how this works out depends on whether the central role in such a semantics is played by truth conditions or by verification conditions. Classical propositional logic, and in particular the law of the excluded middle, flows from a model which has each assertion associated with a certain truth condition which either obtains or fails to. For then the relevant logical constants can be explained by means of the familiar truth tables. For instance, ~ is explained as giving sentences ~p which are true just in case p's truth condition does not obtain, and v is similarly explained as giving sentences p v q which are true just in case either p's truth condition or q's truth condition obtains—from which it follows that p v ~p will always be true. But if, on the other hand, we conceive of assertions as getting meaning from their verification conditions, things work out differently. For then the way we will explain ~ and v (roughly: ~p is verified just in case it is verified that p is not verifiable, and p v q is verified just in case either p is verified or q is verified) will not be such as to guarantee that p v ~p always holds. In particular, statements like '*a* is soluble or it is not the case that *a* is soluble' will not be guaranteed, for it may be that on occasion neither the verification conditions of '*a* is soluble' nor those of 'it is not the case that *a* is soluble' are satisfied.

So the argument is that classical logic, with the law of the excluded middle, emerges from a model of meaning which has a truth condition attached to each sentence; whereas we get something like intuitionistic logic, which does not validate the law of the excluded middle, if meanings are given by associated verification conditions. And so, says the truth conditions theorist, we should be able to tell whether people understand sentences in terms of associated truth conditions or in terms of verification conditions by seeing whether their logical reasoning is classical or intuitionistic. If the community in our earlier thought experiment really did lack any concepts of verification-transcendent truth conditions then they ought to turn out to accept intuitionistic logic. But we are not like that—we do accept the law of the excluded middle, even for those verification-transcendent assertions which are not in practice effectively decidable. So here is a difference between our linguistic practice and the imaginary tribe's, a difference which shows that we do in general understand sentences in terms of truth conditions rather than verification conditions.

Dummett has resisted this line of argument on the grounds that since people can err logically there is no reason why we should take their actual logical practice to show us what kind of contents their assertions have. (Cf. his [1976], pp. 102–3.) That is, Dummett admits that we and other linguistic communities characteristically reason classically about dispositions, say. But he takes this to be a mistake on our and their part. His view is that since we do (have to) understand our assertions in terms of their verification conditions, we ought to reason intuitionistically. In so far as we do reason classically, it is only as a result of our misunderstanding the kind of content our sentences (can) have, and as such proves nothing about what contents they actually do have.

But is this line of argument plausible? So far in this section I have been conforming to the standard literature by attending to the verificationist *'under*definition' of certain terms—to the circumstance that for many assertions it is possible that neither they nor their negations are verifiable. It is this circumstance that has eventually led to Dummett's doubting our entitlement to the law of the excluded middle. But suppose we turn instead to the way terms are also characteristically *overdefined vis-à-vis* their verification conditions—to the fact that for many assertions what we take to be ways of verifying them can, and often do, lead to the simultaneous 'verification' of both them and their negations. What then comes into question is not the law of the excluded middle but the principle of non-contradiction! The truth conditions theorist will point out that, in the case of conflicts of the kind in question, we of course stand by the principle of non-contradiction and diagnose that one or the other of the conflicting assertions must be wrong. And, as before, he will take this to show that the contents we attach to such assertions must go beyond what we currently take to be ways of verifying them (which, after all, make them both right) to some grasp of what makes them true. If Dummett is to resist this argument in the same way as he resisted the appeal to the law of the excluded middle then it seems that he ought to say that it is an error on our part to uphold an unrestricted principle of non-contradiction—that since we (must) understand assertions in terms of their verification conditions we have no warrant for ruling out the possibility that both an assertion and its negation might hold good. But this would surely be absurd, whatever we might initially have thought about the corresponding story about excluded middles. At least we had the reasonably well-understood intuitionistic alternative to classical logic to lend weight to the suggestion that we ought to dispense with the law of the excluded middle. But whatever are we to do if we give up non-contradiction? It is difficult to imagine an analogous alternative logic which might make sense of this suggestion.

In Chapter 6 I shall return to this argument for a truth conditions approach to meaning and defend it in far greater detail. For the moment I shall turn instead to the more immediate question of how much a truth conditions approach as now understood will actually help with the meaning variance thesis. Even if we do allow that inferential practice shows there to be a sense in which we and others understand assertions in terms of truth conditions, we still need to ask whether the same sentence actually will normally have the same truth conditions as used in different theories. And once we do ask this question it becomes clear that there is nothing in the discussion so far that will enable the truth conditions approach to fulfil its promise of showing how inter-theoretical constancy of meaning is possible. The trouble is that no alternative has been suggested for identifying the truth condition of an assertion except via its verification conditions. We can allow that 'solubility', 'bravery', etc. give rise to assertions which need to be understood as about something over and above the recognizable displays which are accredited as verifying those assertions. But as to what that supposed something is, what further answer can we give except to say that it is that property which manifests itself in the accredited displays? And if this is all there is to be said, then what basis could we have for maintaining, of two assertions with different accredited verification conditions, that they are to be understood as having the same truth condition?

Perhaps it should have been clear from the outset that the switch to truth conditions would in the end be of little help with the difficulties raised by the apparently inevitable variation of verification conditions with theoretical change. Consider terms like 'solubility' and 'fragility', or 'bravery' and 'cowardice'. We have now agreed that such terms are to be understood as standing for something beyond their accredited displays. But in this respect they are all exactly the same—they are alike understood as standing for *some* such thing. What that something is, what makes these terms have different contents, is solely that the accredited displays via which that something is identified are different in each case. So the criteria for identifying and distinguishing truth conditions in effect collapse into those for identifying and distinguishing verification conditions, and naturally enough such problems as we had with the latter remain with us.

Nor, obviously, will it help here to focus on reference rather than sense. If the meaning (sense) of a term consists of a concept of something which transcends its overt displays, then the extension of such a term will, if anything, be the class of entities which have that certain something. But if all that can be said about what that something is, is that it is something evidenced in certain displays, then the extension of a term will still be fixed by reference to those displays (verification conditions). And so we will be left with our earlier doubts about the determinacy of extensions; for there will normally be a number of different displays (verification conditions) for any term, and we will as before get different extensions as we concentrate on different displays.

No doubt part of what is involved in construing certain assertions in terms of concepts of possibly unrecognizable truth conditions is the notion that there is some enduring structural basis for the displays which actually enable us to verify those assertions. Thus we might take it that 'solubility' and 'fragility' are at bottom matters of microscopic physical make-up, and that 'bravery' and 'cowardice' depend essentially on certain physiological brain characteristics. This notion raises a number of further issues which will receive full treatment in Chapter 5, when I discuss certain views on related matters put forward by Hilary Putnam and Saul Kripke. But in itself this notion makes no difference to the arguments of this section. Even if we think of 'solubility' or 'fragility' as having some enduring structural basis, the distinctive contents of these terms will still derive from the displays by which we recognize the applicability of these terms; to think merely of there being some underlying basis does not tell us anything new about what that basis is. Nor is the situation altered even when currently accepted theory does contain claims about what the structural basis in question is, about what kinds of molecular structure, say, ensure 'solubility'. The concept of 'solubility' will then indeed be centrally a concept of the possession of such an underlying structure. But this latter concept will in turn still have to be identified via what accepted theory says about that structure and the entities comprising it, via what accepted theory leads us to accredit as satisfactory evidential grounds for verifying claims attributing that structure to something. And so once more it seems to turn out that the meanings of our terms (our concepts) will always vary as the theory in which they are used varies.

4

THE METHODOLOGY OF SCIENTIFIC RESEARCH

1 THE METHODOLOGY OF SCIENTIFIC RESEARCH PROGRAMMES

How are we to proceed at this point? We have been unable to develop any satisfactory way of approaching the paradox of meaning variance. Should we now simply resign ourselves to agnosticism about the objectivity of scientific practice?

One alternative strategy would be to stop talking about meanings altogether. A number of philosophers of science (with Karl Popper the most prominent amongst them) maintain that questions of meaning are unimportant to the analysis of scientific practice. So let us see how far we can get without any mention of meanings.

It will be useful to introduce the discussion by considering Imre Lakatos's 'methodology of scientific research programmes' (Lakatos [1970]). As befits a disciple of Popper, Lakatos does not base his analysis of science on any considerations involving meaning. But he does recognize that such arguments as those of Kuhn and Feyerabend pose problems for 'dogmatic' or 'naïve' versions of Popperian falsificationism. Lakatos's 'sophisticated methodological falsificationism' is designed to provide an objectivist alternative to what he sees as the unacceptable irrationalism of Kuhn and Feyerabend.

Lakatos's central notion is that of a *research programme*. A research programme is an evolving succession of theories. Such a research programme is characterized by a *negative heuristic* and a *positive heuristic*. The *negative* heuristic of the programme derives from a methodological decision to regard certain fundamental theoretical assumptions as effectively irrefutable. These fundamental assumptions form the *hard core* of the programme, and their inviolability means they are common to each of the evolving succession of theories that make up the programme. Each of these succeeding theories is a *version* or *variant* of the programme. They differ in the differing sets of *auxiliary hypotheses* that they conjoin to the fundamental assumptions of the hard core. Such auxiliary assumptions will in general serve to 'pin down' the entities discussed in the hard core in such a way as to make it possible to derive specific predictions from a particular version of the programme. If such a prediction is not borne out then we have an *anomaly*. But this will reflect in the first instance not on the research programme as a whole but only on the particular version which gave rise to the confuted prediction. Because the hard core has been rendered irrefutable by methodological fiat, it will always be the auxiliary hypotheses peculiar to that version that are replaced rather than the fundamental assumptions which define the programme itself. Thus the programme generates a succession of variations on the common theme of the hard core, each successive version producing adjusted sets of auxiliary hypotheses to replace those of the previous version. The *positive* heuristic of a programme 'consists of a partially articulated set of suggestions or hints on how to change, develop the "refutable variants" of the research programme, how to modify, sophisticate, the "refutable" protective belt' (Lakatos [1970], p. 135). That is, the positive heuristic guides scientists when they want to move from one version of a research programme to a later improved one—it tells them what kinds of modifications in their protective belt of auxiliary hypotheses are appropriate. Often this will be done by adopting some kind of 'model' of the mechanism behind the phenomena under study, such as the wave theory of light, or the planetary model of the atom. Or again, it might involve the acceptance of syntactically 'metaphysical' all-some statements, such as 'All changes of motion can be explained in terms of impacts between hard atoms', or 'There is a bacterium to explain every infectious disease.'

Let us illustrate Lakatos's ideas briefly by outlining the elements of the Newtonian research programme ('possibly the most successful research programme ever', [1970], p. 133). The hard core of the programme comprised the three laws of dynamics and the law of gravitation. The auxiliary hypotheses of successive versions consisted of further assumptions concerning such things as the constant of gravitation, the effects of air resistance, the sizes and masses and positions of the planets, and the workings of various experimental instruments. The positive heuristic of the programme then told Newtonian scientists what kind of further suggestions to try when their auxiliary hypotheses required further elaboration. Thus for instance there was the all-some principle that 'All the motions of the solar system can be explained solely by reference to the gravitational interactions of its members.' This directed scientists faced by anomalous planetary motions not to postulate any non-gravitational influences on planetary motions, but instead to check for overlooked further planets, or to re-evaluate their computations of masses, or to make an allowance for the oblateness of the sun, ...

There are obvious affinities between Lakatos's view of science and Kuhn's, between the Lakatosian notion of a *research programme* and the Kuhnian idea of a *paradigm*. Both involve commitments to certain assumptions as effectively inviolable, and both are filled out with the help of 'metaphysical' presuppositions which suggest to researchers the appropriate way to go about solving anomalies or puzzles. What is more, Lakatos sides with Kuhn in emphasizing the impossibility of any easy empirical 'disproof' of a research programme (paradigm)—counter-examples can always be considered to discredit the current auxiliary hypotheses rather than the continuing research programme itself. Nevertheless there are differences. For one thing, Lakatos does not see it as essential to science that competition on fundamentals be eliminated. He allows that a number of different research programmes can proceed side by side in a given area without 'normal' scientific progress being disrupted. And secondly, and rather more importantly for our purposes, Lakatos, unlike Kuhn, has from the start been concerned to show that there *are* principles by reference to which one research programme can in time, if not immediately, be shown to be objectively superior to another.

The key idea here is that of the *progressiveness* of a research programme. Consider again the situation where a particular version of a research programme is faced by an anomaly—where one of the predictions yielded by that version is not borne out. One way of dealing with the anomaly would be simply to abandon one or more of the auxiliary hypotheses involved in the reasoning behind the prediction. Thus we might reject certain instruments as simply being unreliable measures of what they were supposed to measure, on finding that they register results other than those we theoretically anticipated. But such a step would be *theoretically ad hoc* (or *theoretically degenerate*) in that it deals with the problem only by decreasing the empirical content of the research programme, only by reducing the range of empirical facts which the programme is attempting to accommodate. Lakatos says that scientists are required to make *theoretically progressive* modifications of their programmes: instead of just abandoning auxiliary hypotheses in the face of difficulties they should always strive to replace them or adjust them in such a way that the modified programme gives rise to new predictions. Thus we might suggest that certain hitherto unconsidered factors are interfering with the behaviour of the entities being measured; or that the instrument *is* wrong, but that we can get the right results by applying certain correction factors; or we might even devise a new and more sophisticated type of instrument. . . . Each of these ploys will give rise to new and independently testable predictions (that other tests for the interfering factors will have positive results; that the new instrument readings, or the corrected old ones, will work in other situations too . . .). A programme is *empirically progressive* as well as theoretically progressive in so far as some of the new predictions are actually borne out.

Lakatos maintains that scientists are objectively justified in sticking to a research programme only as long as it remains progressive. To be more specific, he requires that successive versions of the programme should manifest theoretical progress, and that this theoretical progress should on occasion be substantiated by the empirically progressive verification of novel predictions. These requirements are in fact fairly moderate. Lakatos does not ask that the programme always prove empirically progressive: occasional successes with novel predictions, even amongst a larger number of failures, are quite enough to justify confidence that the programme is still on the right track. Nor does he even insist that a programme always come up immediately with theoretically progressive solutions to all the problems it faces. There will always be many such problems, and so it will be perfectly proper to shelve some temporarily in an *ad hoc* way, provided only that they are eventually dealt with when time and resources allow (cf. [1970], p. 182). But even taking this moderation into account Lakatos feels he has identified principles which allow for objective decisions between competing scientific views.

Does the meaning-independent account of scientific objectivity promised by the 'methodology of scientific research programmes' really succeed in circumventing the relativist threat posed by the meaning variance thesis? In the end I shall argue that it is certainly of some help. But first it is necessary to deal with certain possible criticisms of Lakatos's ideas. I shall discuss in turn the status of observations, the identity criteria for research programmes, and the vindication of methodologies.

2 OBSERVATIONS AGAIN

One of the arguments for doubting scientific objectivity—indeed the crucial argument for doing so—was that the meanings of scientific terms did not seem determinately tied to elements of sensory experience. This made it unclear what would be wrong with scientists always adjusting the sensory content of their terms in just the way required to retain favoured assumptions in the face of awkward evidence. And this problem went all the way down, in the sense that it seemed to apply to terms characteristically used in making observation reports as much as to abstract terms usually used inferentially in the higher reaches of theory.

Now the methodology of scientific research programmes can be construed as a kind of response to the possibility of scientists always readjusting the experiential import of their terms so as to save favoured postulates: scientists are allowed to so maintain favoured postulates only as long as the revisions of auxiliary hypotheses this requires can be made progressively. However it does seem as if it is being taken for granted here that there is *some* level of scientific discourse at which meanings are not theory-dependent. For 'progress' is defined in terms of the development of sets of hypotheses which will lead to successful predictions. If this is to be a serious standard there surely needs to be some independent set of 'facts' by reference to which predictions can be deemed to succeed or fail. Presumably these facts are to be supplied by observation reports—certainly there is nothing in what Lakatos says to suggest otherwise. However the theory-dependence of observation reports renders them quite unsuitable for this role. If our theories can on occasion quite legitimately correct a previous tendency to respond to certain observational stimuli with certain words, then there is the possibility that it might not be confuted prediction that is at fault but the confuting observation. But if observations are so overturnable then what becomes of the requirement that programmes proceed progressively—progressive with respect to what?

Lakatos himself says little about this problem as such. But some of his remarks, and parts of my earlier comments on observations, suggest the following way of dealing with the problem (cf. [1970], pp. 127–31). Why not conceive of human observers themselves as certain kinds of scientific *instruments*? There is nothing particularly outlandish about this notion. We do suppose that people get trained to issue in various utterances in response to (prompting and) the reception of certain sorts of sensory information. And indeed when we credit an observation report it is surely precisely because we do make this kind of supposition, because we do hold that whenever a suitably disposed and situated human observer issues the utterance in question it is as a result of his interaction with the kind of entity the utterance describes. For instance, if we did not believe that in standard conditions English speakers characterize things directly as 'red' only when red things are present, then clearly we would not credit their observational uses of this term; and similarly, if even more obviously, with trained scientists' observation reports about particle interactions, distinctive accents, etc. Often we will have a fairly sophisticated set of assumptions about how the entities being observed relate, perhaps via instruments, to the sensory systems of appropriate observers. And then we will be able to give some kind of explanation of why and how it is that those observers only utter the relevant words in the presence of those entities. But even in the absence of some such articulated explanatory account we will still at least assume that the observers are somehow so constituted as to make those reports reliable.

If we think of human observation in this way then we can apply Lakatos's methodological principles to the assessment of observation reports themselves. Suppose a theorist wants to challenge a prima-facie acceptable observation report because it confutes some prediction implied by his theory. Our worry was that he would be able to save his theory by simply dismissing the observation report. But now we see that this will also involve him in rejecting at least one previously accepted auxiliary hypothesis about the general connection between human observers' utterances and the presence of the entities putatively observed. And so it can be required that if he is to do this he must be able to do it progressively: he must be able to replace the hypothesis or hypotheses in question by alternatives which lead to new predictions. It is not enough for him to dismiss the human observer in question as unreliable; he must be specific about when such observers are reliable and when not, and he must put forward some definite postulate about what the reports in question do indicate if it is not the state of affairs originally supposed. Normally we might do this via some modification of our understanding of how the sense organs and instruments involved work, thus explaining how a malfunction can give rise to illusions; but again it is not essential we have such a detailed explanation, provided at least we make some definite claim about the new import of the observation reports.

This now seems to plug the gap in the Lakatosian methodology. There are after all constraints on a scientist's entitlement to reject observation reports when they raise problems for his programme. Rejecting such reports itself involves revising the auxiliary hypotheses of his programme, and, as with all such revisions, there is an obligation that they be progressive if the programme is to be retained.

To those brought up in a traditional epistemological environment this might all seem a bit strange. Surely our acceptance of observation reports cannot depend on our having scientific generalizations showing how the occurrence of such observation reports are reliable indicators of the entities putatively reported. The traditional response to any such suggestion is to complain that we could not have come by such generalizations in the first place unless there were some observations we could trust on their own. But this complaint no longer carries any weight. Science does not, as traditionally supposed, proceed by reasoning inductively to generalizations framed in an authoritative observation language. As we have seen, scientific generalizations are formulated rather in terms only loosely related to the making of observations—and of necessity tend to be adopted only on a promissory basis. The role of observation is not to provide a firm basis from which generalizations can then be inductively extrapolated but, if anything, to provide some check on whether the promise of previously made theoretical commitments has been fulfilled. And it is no essential bar to observation playing this role that the significance we attach to observation reports should depend on our having certain generalizations about human observers. Provided the scientist does not have *carte blanche* to discard such generalizations at will, the check provided by observation reports so construed will be real enough. (See Quine [1969], for other arguments in defence of a 'naturalized' understanding of observation.)

A few further remarks about observation will be useful here. So far I have been fairly loose about distinguishing between an observation report considered as a *statement*, to be accepted or rejected as a meaningful claim about reality, and an observation report considered as a 'dumb' *occurrence*, susceptible of scientific explanation and from which inferences about other occurrences might be drawn. What I am in effect arguing is that, if we accept an observation report as a statement, it is because we accept a generalization to the effect that whenever such reports occur then the state of affairs they report (*qua* statement) obtains. (Why ever should we accept an observation report otherwise?) Now, when a scientist suggests that a certain observation report ought to be rejected, what he is doing is questioning it *qua* statement (and correspondingly questioning the generalization which hitherto validated such reports). The constraint on his doing this wantonly can be considered to derive from the fact that such questioning leaves untouched the status of the observation report *qua* occurrence. And as such the observation report remains in need of explanation. The obvious explanation would have been that the report occurred because of the observer's interaction with the state of affairs putatively reported. But this explanation is no longer available, for the questioner has denied the generalization which substantiated it. So the questioner is obliged to be 'progressive' and produce some new general hypothesis by which to explain the occurrence of the observation report in question.

Put like this, there might seem to be a kind of regress threatening. The questioner owes us something further by way of explanation only in so far as he continues to accept that the observation report did at least occur, even if he no longer accepts it as a statement. But why should he accept even this? Suppose the observation report in question was 'Here's something red' said by X on some occasion. Why should Y, our doubter, not only deny that there was something red there but also deny that X's observation report even occurred? If he did this then presumably his remaining problem about explaining why that report occurred would disappear. But note that this further step involves rejecting another observation report, namely 'X said "Here's something red" '. And even if this is rejected *qua statement*, there once more remains the *occurrence* of that report to be explained. Y can legitimately deny that the first observation report ('Here's something red') was made; but if he does so he still needs to come up with some new explanation of why he and anybody else in observationally appropriate conditions should have been inclined in the first place to report 'X said "Here's something red" ', if X didn't. There is nothing in principle to stop Y going down this regressive path. But given that he will have to keep on producing alternative accounts of our most basic and widely-used observational abilities, he will find the going increasingly heavy.

Of course Y could always avoid his problems by closing his eyes and ears; or he could stop argument by pretending to have done so. But we should scarcely expect a methodology of science to be proof against uninquisitiveness or duplicity.

Thus we see how observations can provide a check on theories without constituting an immovable sounding board of indubitable 'facts'. Judgements about particular situations can either consist of a direct observation report, or can come indirectly from inferences from such reports. Direct observation reports are no less revisable than indirectly inferred statements about particular situations. But, as in the latter case, there are constraints on such revisions. An inferred statement about a particular situation gets its warrant from the generalizations behind the inferences involved. So a scientist questioning such a judgment, as he might in order to save certain favoured postulates, will have to question one or more of those backing generalizations. The Lakatosian methodology specifies that if he is to do this he is obliged to find some replacement for those generalizations, otherwise his theory will simply become weaker, and he will end up with no idea of what to infer from judgements he was previously able to make something of. Accepted observation reports do not get their warrant from inferences, but in a sense just from being made. But still, once such an observation report is questioned then attention is drawn to one or more implicit generalizations which its previous acceptance committed us to. And so, once more, such questionings are constrained by the obligation to find replacements for those generalizations. However, despite this constraint, there is no limit in principle to such questionings—as we have seen there is a bottomless sequence of observation reports, and observation reports of such reports, . . . , each coming into consideration if we manage to reject the previous one.

A related point. It would be a mistake to understand the arguments of this section as suggesting that the sole task (and test) of all scientific theories is to explain the occurrence of observation reports. Far from it. Observation reports are only a very small part of what occurs, and not a very interesting part at that. The task of science is to come up with theories that will explain everything that happens or as much of that as possible. Science is not even restricted to explaining the occurrences *reported in* observation reports (including those that are about things other than other observation reports)—for many particular occurrences are found out about by inference rather than direct observation. But, still, observation reports, and what they (if acceptable) report on, are some part of what happens, and so science ought if possible to explain them along with everything else.

3 THE INDIVIDUATION OF RESEARCH PROGRAMMES

Lakatos's notion of a research programme depends crucially on the idea of a 'negative heuristic' which directs scientists not to meddle with their 'hard core' (and, to a somewhat lesser extent, on the 'positive heuristic' which suggests how they might go about developing new sets of auxiliary hypotheses). What warrant is there for supposing that scientists do actually commit themselves to hard cores in the way supposed? Lakatos suggests that the adoption of a certain hard core is a matter of methodological decision, of the adoption of certain methodological rules (Lakatos [1970], pp. 132–4). But he does not in fact present any evidence to show that scientists do periodically pick out certain fundamental assumptions and commit themselves to not modifying them for the foreseeable future. Indeed, once we think about it, it seems most unlikely that there is any such evidence to be found.

In a way the suggestion that scientists work in programmes with privileged hard cores is reminiscent of the analytic-synthetic distinction—and it is susceptible of much the same criticisms. If scientists do not overtly show their special attachment to their supposed hard cores in any way, then what reason is there to think that they actually do form any such attachments?

We can bring out the problems involved here by considering a historical or sociological investigation of, say, mid-nineteenth-century chemistry, or post-war twentieth-century genetics. How is the historian or sociologist to tell how many research programmes were being pursued in the scientific community in question? Presumably there will be some assumptions common to all the scientists concerned—but then there will be others on which they held variously different views. What is to be made of such differences? Were they just differences within a given research programme, just a matter of different scientists upholding different versions of a single programme—or do they actually indicate different programmes? And, if the latter, were there two programmes in competition, or three, or what? If we could tell whether the contentious assumptions were parts of hard cores or whether they were merely auxiliary hypotheses then clearly we could give an appropriate answer. But without some further explanation of this distinction we seem stuck.

Some remarks in Kuhn's [1970a] are addressed to this kind of problem. (Kuhn faces essentially the same difficulty as Lakatos—he needs to explain exactly which commitments are constitutive of allegiance to a given paradigm, and which can be varied without abandoning the paradigm.) Kuhn suggests that we can start by identifying 'scientific communities' externally, as groups which subscribe to the same journals, attend the same conferences, circulate typescripts to one another, etc. Then a paradigm can be identified as the set of assumptions and principles held in common by all members of such a 'community'. But this is scarcely an acceptable solution to the problem at issue. For a start, one might query whether a notion of 'paradigm' (or 'research programme') which was conceptually predetermined to come out one per so-defined 'scientific community' was the most useful tool for analysing science. (Lakatos, for one, would presumably wish to allow that there could be more than one 'research programme' or 'paradigm' in such a community.) Apart from that, there is the point already made in connection with the analytic-synthetic distinction, that a distinction between postulates, the sole substance of which derives from what a group of scientists *do* agree on, will not be something which *'accounts* for the relative fulness of their professional communication and the relative unanimity of their professional judgements' (Kuhn [1970a], p. 182, my italics). And indeed we are even worse off here than we were with the analytic-synthetic distinction. For presumably it happens on occasion that the members of a 'scientific community' actually cease to agree on the constitutive elements of a paradigm (or a hard core), as when there is a 'scientific revolution' (or a change of research programme). But if this is right then the idea of identifying the constitutive elements of a paradigm or a research programme by reference to what a 'scientific community' agrees on collapses completely. It is no good suggesting that it is what a scientific community agrees on specifically during periods of normal science (while merely elaborating and refining auxiliary hypotheses) that counts. For in the absence of a specification of what is to count as a paradigm (research programme) how are we to tell in the first place whether some scientific development is a solution to a normal scientific puzzle (change of auxiliary hypotheses) or a revolution (change of hard core)?

Once questions such as these are raised it is difficult not to be sceptical about finding answers. Did the realization that gases of elements have diatomic molecules and the consequent adoption of

Avogadro's hypothesis amount to a scientific revolution, or was it merely a solution to a puzzle within the existing paradigm of nineteenth-century chemistry? Did the discovery of DNA institute a new research programme for modern geneticists, or was it just a matter of new auxiliary hypotheses? There is little in either Kuhn or Lakatos to allay the feeling that these questions are somehow misconceived.

(It might be thought that the notion of a 'positive heuristic' could help to individuate research programmes and so explain what counts as a switch from one such programme to another. But essentially the same arguments apply. Scientists do indeed adopt models, metaphysical principles, etc., and these do play an important part in directing the development of their research. But of course there will again always be differences on such matters, both across individuals and over time. And so we cannot start thinking seriously in terms of '*the* positive heuristic of *a* research programme', without immediately being faced with unanswerable questions about which differences in heuristic ideas indicate different programmes and which do not.)

So whichever way we look at it, the notion of a scientific research programme seems to lack substance. On the other hand, there is surely some sense in which scientists are more committed to certain assumptions than to others. As Lakatos observes, when Newtonian theorists discovered that the orbits of certain planets were not as theoretically anticipated they did not therewith abandon the laws of motion and gravitation—instead they stuck by those laws and used them to work backwards from the anomaly to find out which of their other assumptions could be faulty (Lakatos [1970], pp. 100–1). Lakatos sees it as one of the merits of his account of science that it captures the continuity of scientific research ([1970], p. 132). On, say, a simple Popperian view, a 'theory' which is faced with a counter-example is abandoned and replaced by a different 'theory'. But to say this and nothing more is to obscure the important fact that the new theory will invariably be largely continuous with what went before, will invariably be a modification of the old theory. It is not as if the history of each scientific field consisted of a succession of falsifiable but not then yet falsified hypotheses, each completely unrelated to what went before and what came after. Surely what we find is the development, elaboration, refinement of frameworks of basic assumptions which are retained through time and therefore make for continuities between succeeding views. The problem is to reconcile this insight with the implausibility of supposing that certain postulates are granted a qualitatively distinct status by some mythical methodological fiat.

Fortunately there is an acceptable substitute for the notion of a 'hard core' which will allow us to retain what is valuable in Lakatos's methodological ideas. Recall the remarks on the deductive systematization of theories made at the end of Chapter 1. In particular recall the observation that certain of the generalizations in such a deductive systematization will be more *central* than others: some will function more often than others as premises in the deduction of derived generalizations. (Thus the second law of motion was argued to have been more central in classical mechanics than, say, Hooke's law.)

Now consider a group of scientists working in a given area. We would expect them all to differ in at least some respects in their assumptions about their subject matter. But nevertheless we can use the notion of centrality to chart the structure of their various theoretical affiliations. At some level of centrality they will almost certainly all share assumptions—even if only at the level of such principles as the conservation of energy or the uninheritability of acquired characteristics. On assumptions at a somewhat lesser degree of centrality they will split into a small number of camps. On still less central assumptions there will be further fragmentation, until when even the least central generalizations are taken into account it is unlikely that any two scientists will agree entirely. We can picture their various theoretical commitments as forming a kind of tree. The trunk consists of those central assumptions common to all scientists in the field. The first branching of the trunk corresponds to those more basic points of initial disagreement which divide the community into a small number of groups. And so on with further branchings, until we get to the thinnest twigs, to individual disagreements on the least central assumptions.

This metaphor should not be confused with the pyramid metaphor used earlier. *Each* path taking us from the trunk of such a tree to one of the outermost twigs corresponds to a *whole* pyramid. For such a path through our tree indicates the total set of assumptions making up a given scientist's theory: it gives us first his most central assumptions, the ones he shares with all scientists in the field; then the rather less central ones he shares with some but not all of his colleagues; and so on, down to his least central and most individual views.

The point of introducing this tree structure is that it enables us to formulate a modified version of Lakatos's methodological analysis. Suppose we consider any initial sequence of trunk, branch, branch, . . . as defining a *line of research*. That is, a line of research is any initial segment common to a certain number of total paths (theories) going from trunk to twig. Then any scientist will turn out to adhere to a sequence of related lines of research, each of increasing specificity. At the most basic level all scientists in the area will be pursuing the same line of research, for they will all be working within the common assumptions comprising the trunk. At less basic levels they are following different lines of research, the more so the less central are the assumptions we take into account.

Now, lines of research, as so defined, will provide, at whatever level, the kind of continuities in scientific development required for the application of Lakatosian methodological precepts. It is not difficult to see why this should be so. The task of replacing a rejected assumption—as must be done if the generality of the theory is to be restored, if the theory is not to be allowed simply to degenerate—will be the more arduous the more central that rejected assumption is. For if a new assumption in a deductive systematization is to be made to tally with the evidence of observation reports, then it will be necessary to revise correspondingly the other assumptions in the deductive systematization which together with it play a part in implying derived generalizations formulated in the terms used in those reports. The more central the new assumption the more intellectual effort will be required to produce these revisions, and the more time and resources will be required to assess observationally such hitherto unchecked derived generalizations as will inevitably arise from those revisions. (Such revisions might actually lead to the rejection of certain observation reports, in the way described in the last section. But the same points still apply, for as we saw such rejections will always leave the scientist with other observation reports that need accommodating.)

So it is natural enough that scientists should in the first instance always take the line of least resistance and seek to modify their least central assumptions first. Which means that the notion of 'centrality' can function as an acceptable substitute for the Lakatosian concept of a 'hard core'—the preferential disinclination of scientists to modify their more central assumptions shows why there should be continuity in the development of scientific theories. It is true that we get a rather different representation of a scientific community's ongoing affiliations: instead of one or more discrete 'research programmes' we get a complex structure of branching 'lines of research'. But, as should now be clear, Lakatosian principles can still be applied perfectly straight-forwardly to the more complex continuities given by lines of research. At any level, a group of scientists pursuing a given line of research will be obliged to do so progressively. They will be entitled to retain their common set of central assumptions only as long as they can continue to put forward new auxiliary (less central) hypotheses which increase the range of derived generalizations in their theories, and if, in addition, some of the new predictions to which those new hypotheses give rise are borne out. If and when they can no longer do this, then it will be time for them to switch to a different line of research (at that level), to retreat to a lower branching point and look for a different path back up the tree.

It is worth spelling out the advantages of 'lines of research' over 'research programmes'. In the first place it is important that there is no need in my modified account for any *methodological decisions* on the part of scientists. The 'more/less central' continuum on which the definition of 'lines of research' depends is not, implausibly, set up by any conventional fiat on the part of scientists. Scientists do not *decide* that they are going to be quicker in revising some assumptions than others. On the contrary, it is an inevitable feature of the actual structure of their scientific theories that certain of their assumptions will *be* more central than others; and it is a natural result of this difference that they will be less ready to meddle with more central assumptions than with less central ones. Given that scientists want their ideas to progress as quickly as possible, they do not need any extra constraints to make them retain as many of their central assump-tions as possible—for as explained, that centrality itself means that if there is any short route to progress it will be found by leaving those assumptions intact.

A further advantage of the 'lines of research' approach is that it yields an even better picture of the continuity of scientific develop-ment than the description in terms of 'research programmes'. I said above that a merit of Lakatos's account was that it corrected any naïve falsificationist tendency to suppose that all changes of theory were matters of hypotheses being replaced by completely unrelated ones. But on the level of changes of programmes the original Lakatosian account itself reimposes an entirely analogous conception: it suggests precisely that we do sometimes get discontinuities of a distinctively complete kind, namely when a switch is made from one hard core to another. Nothing is said about the possibility of greater or lesser agreement between the different hard cores in question. However, the original criticisms of the naïve falsificationist picture apply equally well even to the more sophisticated Lakatosian imputation of occasional high-level discontinuities. Surely something is preserved in even the most fundamental (or 'revolutionary') of scientific developments? Such changes may indeed require much reworking of previously accepted lower-level assumptions, as explained, and (if we were talking in such terms) we may therefore want to say they involve wholesale changes of meanings or concepts. But considering successive theories just as structures of interrelated assumptions, will we not always find that *some* of the fundamental components of such structures are retained through theoretical change? Even relativity physics contains an assumption corresponding to the classical F = ma (even if many of the further assumptions involving the terms therein are quite different).

All this comes out quite happily in terms of 'lines of research'. The picture we then get is not one of continuous developments within programmes interspersed by occasional and qualitatively quite different switches between programmes. Instead we have an always changing tree of branching theoretical commitments. New twigs and branches are constantly growing as scientists adopt new lines of research at different levels—and other sections of the tree wither and die, as scientists drop old and unsuccessful lines of research. Some of these changes will obviously be more important than others, will involve thicker, more central branches. Sometimes even part of the trunk will fall away and be replaced, when assumptions hitherto unanimously agreed on are discarded and alternatives are thought up. But, notwithstanding the different levels at which such switches of affiliation can occur, there is no reason to suppose that some are different in kind from the others, that some are distinctively discontinuous. For all such theoretical changes will retain something of what went before, will retain something of what was previously accepted by those involved in the switch.

4 THE VINDICATION OF METHODOLOGIES

So far I have been suggesting that with suitable modifications Lakatos's methodology of scientific research programmes shows how objective decisions between competing scientific views are possible.

But what shows that Lakatos is actually right about science? What shows that the proper way to conduct science is indeed to have scientists variously pursuing competing lines of research, with survival of such lines depending on an appropriate rate of 'progress'?

It is not as if Lakatos's precepts were uncontentious. Traditional Popperians, for instance, would insist that scientific seriousness requires a scientist to specify beforehand, for any hypothesis he accepts, exactly what observations would lead him to give it up. They would consider it the antithesis of proper practice for a scientist to try to save a hypothesis in the face of unexpected observations by repeatedly suggesting alternative accounts of how those observations relate to that hypothesis. Kuhnians would disagree with Lakatos about the value of competition: for them progress is best achieved by maximizing agreement on basics rather than by allowing fundamentally different lines of research to proceed side by side. And then there is Feyerabend, who in his most recent writings has been explicitly recommending 'epistemological anarchism' on the grounds that any attempt to lay down rules for rejecting scientific views will only repress scientific advance.

How are such disagreements to be adjudicated? It is natural enough to turn here to the history of science. Thus Lakatos's writings include a number of detailed case studies purporting to show how the programmes instituted by such figures as Copernicus, Prout, Bohr and others indeed involved the more or less progressive elaboration of successive sets of auxiliary hypotheses around a given hard core (Lakatos [1970]; Lakatos and Zahar [1976]). Similarly Kuhn appeals to the patent absence of progress in the pre-paradigm phases of various disciplines to show the value of agreement on fundamentals (Kuhn [1962], Ch. II; [1970b], Section 3). And Feyerabend argues in convincing detail that the theoretical innovations due to Galileo were possible only because Galileo ignored every conceivable methodological precept (Feyerabend [1975], Chs. 6–11).

These kinds of arguments present something of a puzzle. Although there are disagreements about the actual analysis of particular historical examples, all parties seem to recognize the appeal to history as in itself cogent and relevant. If Galileo's practice were indeed at variance with Lakatos's picture of science, then surely this would do something to cast some doubt on Lakatos's methodology. But why should it? Why should Lakatos's precepts have to fit Galileo's practice? That is, why should Lakatos not simply decide that Galileo's practice was unscientific? Not everybody who calls himself a scientist merits the title. And even those who do will on occasion do unscientific things.

It seems as if we ought to be caught in a closed circle here. How can we decide between scientific methodologies by appealing to examples of good scientific practice, when presumably we need such a methodology in the first place to tell us which those are?

Still, the historical arguments in question do seem relevant. This suggests that behind the methodological dispute there must some-how be a kind of implicit agreement about what counts as a successful scientific development, as a scientific advance. Yet, again, how is this possible, given that the substance of the dispute is precisely how science should best be conducted?

In order to resolve this apparent paradox we need to attend to Lakatos's distinction between a *version* of a research programme and the *research programme* itself. A version of a research programme, recall, consisted of a complete set of auxiliary hypotheses conjoined to a hard core. The research programme itself was then the continuing tradition of successive variations on the common theme of that hard core. (When the issues discussed in the last section are of no immedi-ate relevance I shall henceforth switch back to the more convenient if less accurate terminology of 'research programmes', 'hard cores', 'auxiliary hypotheses', etc.)

It seems to me that all parties to the methodological dispute are in at least implicit agreement on what shows the worth of complete sets of interrelated hypotheses, of *versions* of research programmes. Their differences emerge only when we move on to the further question of what shows the worth of a research programme itself, what should lead scientists to continue working with certain central assumptions.

Remember that it is a version, rather than a research programme as such, that can give rise to definite experimentally testable predic-tions. And so it is versions that in the first instance are discredited by failed predictions, by *anomalies*. And in this connection nobody denies that versions *are* so discredited by anomalies. Nobody denies that anomalies have to be dealt with—that a version facing some anomaly should if possible be replaced by some other version which avoids that anomaly. Feyerabend, it is true, makes much of the methodological undesirability of new ideas being discouraged merely because of the existence of apparently contradictory observational 'facts'. But even he does not mean to suggest that such awkward 'facts' should simply be overlooked. Thus Feyerabend shows how Galileo was indeed faced initially by many observationally refutations of the Copernican hypothesis ('stones fall *straight down*, yet do not land to the west of where they are dropped from'). But he does not claim that Galileo ought simply to have turned a blind eye to those 'facts'. On the contrary, it is an important part of Feyerabend's story that Galileo and his followers were, in time, able to reinterpret the anomalous observations so as to make them consistent with the heliocentric system. So Feyerabend, like everybody else, recognizes that anomalies do need to be resolved. (Again, it is of no consequence here that we might do this by revising the assumptions which led us to credit the confuting observation, rather than those behind the confuted prediction.)

Another feature of versions of research programmes which is implicitly but universally recognized as a sign of worth is their relative *generality*. This principle is obviously behind Lakatos's demand that programmes should 'progress', that the replacement of one version by another should increase rather than decrease the empirical detail accommodated. But this principle is in itself by no means peculiar to Lakatos: even those philosophers of science who do not talk of 'research programmes', 'versions' thereof, or 'progress', will scarcely deny that it is always better if possible to add generalizations to a current stock of assumptions rather than subtract them.

So there is at least some degree of covert consensus behind the methodological dispute—there is agreement at least that any total conjunction of interrelated generalizations is the more to be preferred the more it is anomaly-free and general. (Whether there are still further criteria, such as simplicity, say, for evaluating versions of research programmes, and if so whether they are uncontroversially agreed as such, are questions I shall not pursue here.)

It is only when we move on to the conduct of research programmes themselves, to the question of when and whether scientists should stick with certain central tenets in moving from one version to another, that the methodological disagreements arise. For then we are no longer asking about the worth of a total and testable conjunction of generalizations that might be accepted at a given time, but asking how best to *develop* such conjunctions over time. It is then that we face such questions as to whether it is most fruitful for all scientists in a given area to concentrate on one set of basic ideas (Kuhn); as to whether any definite restrictions on the retention of central tenets will threaten the elimination of worthwhile ideas (Feyerabend); etc.

In fact questions on this level presuppose the agreed criteria for evaluating specific versions of programmes. For when we ask how scientific views are 'best' to be developed, the aim with respect to which some methods are to be deemed better than others is precisely, if implicitly, that of ending up with as general and anomaly-free total conjunctions of generalizations as is possible.

In a sense we might say that the simple Popperian requirement that scientific theories ought to be falsifiable applies perfectly straightforwardly when the unit in question is a specific version of a programme. Considered as a whole the conjunction of generalizations making up such a version will be falsifiable, in that it yields definite predictions and is to be rejected if such predictions are not borne out. And such a version is the more satisfactory the 'more falsifiable' it is, the wider the range of empirical detail it accommodates. (Cf. Popper [1959], Ch. VI.)

The methodological disputes arise only when we shift the meaning of 'theory' ('position', 'view', . . .) and consider as the unit to be evaluated some favoured sub-set of central postulates around which successions of auxiliary hypotheses are going to be developed.

When we talk about Newtonian mechanics, the kinetic theory of gases, Mendelian genetics, etc., as *theories* what we are talking about is not a total conjunction of generalizations capable of definite empirical refutation but rather something like a continuing programme of research. This is, I think, the more natural sense of 'theory' (though for expository reasons I shall henceforth use 'theory' unnaturally, to stand for total conjunctions of accepted generalizations, unless I indicate otherwise). So we can understand Kuhn and Feyerabend's initial disquiet about the possibility of objective 'theory' choices in science as due to their realization that significant scientific disputes are not about simply falsifiable Popperian hypotheses, but rather about sets of basic postulates which retain an identity over time as the less central assumptions surrounding them are refined and elaborated. However, in emphasizing the complexity of such choices Feyerabend and Kuhn have obscured the important and basic point that there *is* unproblematic and objective agreement about what the refinement and elaboration of basic postulates is aiming at—namely to come up in the end with general and anomaly-free totalities of accepted generalizations. And Feyerabend and Kuhn do implicitly recognize this point, even if they do not explicitly acknowledge it. For it is what gives their historical arguments the force they do have. When, for instance, Feyerabend appeals to Galileo to expose the dangers in any reverence for established 'facts', what he is trading on is of course our awareness that the tradition instituted by Copernicus and carried on by Galileo was in the end undoubtedly successful in producing detailed and empirically accurate theories.

In any case, now we are clear that the questions in dispute are specifically about the conduct of research programmes, and that there is agreement on the criteria for evaluating versions of such programmes, it is easy to see how the appeal to history can avoid begging those disputed questions. For what the history of science can show us is which general strategies for conducting *research programmes* have in the past actually succeeded best in leading to the elaboration of *versions* satisfying the agreed criteria. That is, we can, as with every such question, assess the alternative suggested *means* to the agreed *end* of developing good versions, on the basis of evidence about which such means have achieved that end in the past. Of course the end in question has to be given independently of what that evidence shows us. But what I have been arguing is precisely that there is an agreed answer to the question of what count as good scientific results. What is not agreed is how to get them. But that is a different question, and, given agreement on the former, there is no reason why history should not indicate an answer to the latter.

(There is another respect in which the appeal to history might appear circular. What history is supposed to show is in effect a *theory* about whether certain practices will produce certain results. But how can we use such a historically evidenced 'theory' to decide questions about the appropriate criteria for choosing 'theories'? If we do not know the answer to the latter questions, then how can we know if the deciding 'theory' is any good in the first place? However the distinction I have been urging also breaks this circle easily enough. What we are trying to decide is what is the right way to conduct programmes. These will involve criteria for choosing 'theories' only if 'theory' is read as 'ongoing programme'. But the 'theory' that will enable us to discover those criteria will not be such a programme but a version of one: for it will be a theory that tells us what to do, that predicts that given certain actions, certain results will follow. And so to validate that theory we need criteria for evaluating versions, not programmes—which is just what I have argued we do already have. We will of course not yet have substantiated any thoughts we may have about how best to arrive at such a theory (version). But, given those we have arrived at, by whatever means, we can still use the uncontentious criteria for evaluating versions to tell us which is the best.)

This is obviously not the place to attempt any adjudication of what the historical evidence actually does suggest to be the best way of conducting continuing scientific research. I myself am inclined to suppose that it supports the (modified) Lakatosian view of things, and shall, where necessary, henceforth simply assume this without further argument. But it will be worthwhile making a few brief remarks here about one related point, namely the time factor implicit in Lakatos's methodology. Feyerabend has pointed out the difficulties inherent in Lakatos's allowing that a research programme should not always be required to display immediate progress: *how much* time is then to be allowed? If we say that progress needs only to be achieved in the indefinite long run, then it seems we end up with no methodology at all—a scientist will always be entitled to stick with his favoured ideas for as long as he likes. If on the other hand we impose some specific time-limit we run the risk of snuffing out worthwhile programmes prematurely—it was after all an extremely long time indeed before the Copernico-Galilean programme managed, with various stops and starts, to progress past its Aristotelian competitor. Feyerabend accuses Lakatos of wanting to have it both ways, of wanting to allow that it is never definitely wrong, albeit 'risky', to stick with a degenerating programme, while at the same time suggesting that scientists and others concerned with the disposition of scientific resources are justified in favouring progressive over degenerating programmes. (Feyerabend [1975], Ch. 16.)

Feyerabend's accusation is a not unfair response to the rather unsatisfactory remarks on this matter to be found in Lakatos's [1971]. But the understanding of the appeal to history we have now reached shows that Lakatos's intuitive desire to have it both ways here is not entirely unreasonable. The history of science is supposed to show us which methodological strategies have succeeded in producing general, anomaly-free conjunctions of generalizations. So why should it not show in particular the length of the period of continued degeneration after which further persistence does not produce successful versions? That is, might we not find that short periods of degeneration up to a certain limit can get followed by success, but that periods of longer degeneration do not? But of course it will not be that simple. For one thing, the length of persistence cannot be the only relevant factor—the quantity and quality of effort involved must also matter.

And then what we will be interested in will not just be 'success' as such, but the amount thereof—*how* general and anomaly-free were the versions generated? (For the choices we are trying to inform will always involve asking whether one programme is (will be) *more* successful than another.) Moreover, any attempt to relate energy expended to success generated will have to take into account the 'depth' of the line of research being persisted with—if a 'central' line of research does eventually pay off, it will characteristically pay off more than a 'non-central' one. And there are no doubt many further complexities. All of which makes it quite obvious that we are not going to find any simple specification of exactly when a scientist will be justified in persisting with a degenerating programme and when he will not.

On the other hand, there is no reason to think we will not be able to find any such indication. Why should it not be possible to extract from history at least general principles of the form that such-and-such a programme is more *likely* to be more successful than that? True, even such statistical claims will only be tenuous, for the evidence will inevitably be fragmentary and complex. But we would still expect to get some idea of what is and what is not likely to work. An analogy might be helpful here. A gold prospector cannot say exactly how much unsuccessful working of a site shows for certain there is no gold to be found there. But this does not mean that his past experience, such as it is, does not give him any idea of when his lack of success indicates that the next site offers better expectations, taking into account the relative sizes of the strikes to be expected from the current site and the next, the degree of care with which he has been looking so far, etc. Of course—and this is where Lakatos's intuition comes from—on any given occasion he might have struck lucky if he had persisted with the first site. As with all decisions under uncertainty the right choice can on particular occasions lead to the wrong result. But this does not stop it being the right choice, in the sense of being the one which experience shows to give the greater expectation of success in general. And in this respect research programmes are just like gold mines—an obstinate persister might always strike lucky. So it is perfectly consistent to allow, with Lakatos, that persisting with a degenerate programme might in particular cases lead to eventual and worthwhile progress, and yet to insist that such persistence is in conflict with the general canons of scientific rationality.

5 HOLISM ABOUT MEANINGS

Is the account of science developed in this chapter so far really, as
suggested, independent of any considerations involving meanings?
There has certainly been very little in the way of *explicit* mentions of
meanings. But it could well be argued that there is *implicit* in this
account a quite definite conception of the meanings of scientific
terms, and moreover one which is in tension with the overt objectiv-
ism of the explicit account.

Dummett has observed that 'there is nothing more that we can
require of a theory of meaning than that it give an account of what it is
that someone knows when he knows a language' (Dummett [1974a],
p. 354). From this general perspective it can well be argued that the
approach adopted in this chapter does indeed present a certain model of
meaning for scientific languages, does indeed specify what a scientist
has to know to know the language of scientific theory—namely
that he has to know nothing more nor less than that theory itself. For it
is precisely because the scientist knows the generalizations making up
a theory that he will be competent to assess other statements made
using the terms in the theory's language. This is obvious in the case of
theory-backed inferences from observation reports, for there it is explicit that the warrant for the conclusion of the inference is an assumption of the theory. But, as we saw in Section 2 above, the point applies
also to direct observation reports themselves, in that the acceptability
of such reports is also in the end answerable to theoretical assumptions.

If this is right, then it seems that the account of science developed
in this chapter implicitly carries with it a *holist* conception of meaning. Dummett has made essentially the same observation about
Quine's view of language. Quine denies that words have meanings in
the sense of being associated with sense impressions, or being embed-
ded in analytic meaning postulates; instead he simply analyses languages in terms of undifferentiated networks of behavioural dispositions to use words. Dummett's observation is that this is not a total
rejection of the notion of meaning, as some have supposed, but rather
an attempt to replace a traditional empiricist view of meaning by a
holist one ([1974a], pp. 354–5). Now, the way scientific practice has
been analysed in this chapter is in various respects highly Quinean
—and so it could correspondingly be argued not to abandon the
notion of meaning altogether so much as to smuggle in an implicitly
holist model of meaning.

I am not entirely happy about accepting the terminology of a 'holist' theory of meaning. The reason is obvious enough. We normally and traditionally think of the meaning of words as somehow specifying what entitles us to accept sentences containing them. (On a verificationist conception of meaning it is quite explicit that meanings are what do this. But even a truth conditions approach is likely, as we have seen, to leave us with a close link between truth conditions and verification procedures.) But then it seems that a 'holist' theory of meaning of the kind being mooted will necessarily preclude any account of what entitles us to accept *generalizations* containing scientific terms. For if meanings depend on all accepted generalizations then the very raising of the question of whether to accept or reject a generalization would seem simultaneously to disrupt entirely any grasp we might have had of how to decide that question.

On the other hand, I am not inclined to quibble about terminology here. Part of what I hope my arguments will eventually show is that there is no possibility, or need, of anything to play quite the traditional role of meanings in respect of scientific generalizations. We have certainly had little success in finding a suitable candidate so far. Not that nothing can be said about how our understanding of scientific generalizations contributes to decisions as to whether to accept or reject them. But it will not be a simple matter of grasping stable meanings which persist through such decisions and thereby indicate how they are to be made. And so, given that nothing is going to carry out this traditional function of meanings, there perhaps remains no strong reason for resisting the attribution of a 'holist' theory of meaning.

A full account of how we are to conceive of decisions on generalizations will have to wait until Chapter 6 and a discussion of the logical form of generalizations. But it will be appropriate here to make some preliminary remarks on the holist view of meaning. It might be difficult to see how the admission of any kind of holist conception of meanings can avoid bringing back all our earlier relativist problems with a vengeance. Must not such a theory collapse into the 'theoretical context' view of meaning discussed in Chapter 2, and was that view not quite rightly dismissed out of hand on the grounds that it implied that any revision of theory could only be an arbitrary decision to change the meanings of words?

However we must not allow any implicit commitments we may have to holism about meanings to obscure the lessons of the earlier sections of this chapter. In particular we have seen that, notwithstanding any such holism, it is of course perfectly possible for evidence to compel revision of a scientific theory, as when the existing theory plus observational interaction with the world gives rise to a confuted prediction. When this happens it is quite uncontroversial, as pointed out, that the conjunction of generalizations facing the anomaly requires remedying.

While unproblematic, this responsiveness of theory to evidence does show something important. If it is indeed right to say that the meanings of scientific terms are constituted holistically by the total context of generalizations surrounding them, then one thing we will have to allow is that there can be rational grounds for changing the meanings of scientific terms.

It is a natural enough consequence of any traditional conception of meaning that changes in the meanings of words can only result from more or less arbitrary and non-rational considerations. What substantial reason could there be for permuting the existing associations between words and their contents? So far I have not questioned this assumption of the irrationality of meaning change. Indeed I in effect traded on it in a number of earlier arguments: the primary objection to construing all accepted generalizations as contributing to the meanings of scientific terms, or all observational procedures as doing so, was that there were clearly sometimes rational grounds for revising accepted generalizations, and for revising observational procedures.

In the earlier context of discussion these arguments were not inappropriate, for the assumption of the irrationality of meaning change was common coin to the parties involved in the meaning variance dispute. Those suspicious of the meaning variance thesis objected that it made scientific change inevitably irrational; while supporters such as Kuhn and Feyerabend inclined conversely to argue that, since meanings did inevitably vary, scientific change could not be construed as straightforwardly rational. But if the way out of the paradox requires us to allow a holist conception of meaning, then clearly it is time to abandon the idea that all meaning changes must be irrational. And indeed, from a more general perspective which abstracts from traditional theories of meaning, there is nothing intrinsically implausible in the notion that an adequate factual account of some area might better be achieved by one system of meanings than another.

Perhaps there remains further reason for an initial disquiet about a

holist conception of meaning. Surely, it might be said, scientific
rationality usually requires more than just that *some* change be made
in a complex of accepted generalizations facing an anomaly: when
such an anomaly occurs is it not normally also clear *which* bit of that
complex needs changing? Surely when unexpected experimental
results occur it will often be quite obvious to the scientists concerned,
and quite rightly so, which of their previously accepted assumptions
has been discredited. But a holist conception of meanings seems to
leave no room for anomalies rationally to compel some revisions rather
than others. For on a holist conception all adjustments of a theoretical
structure in response to anomalous evidence seem on a par, for they
will all equally amount to changes in the meanings of the relevant
words. If any theoretical adjustment alters the existing meanings of
all the relevant terms, then how can those meanings indicate one
adjustment rather than another as the appropriate one? But then what
is it that stops scientists with a given theory always responding to
anomalies in different ways—some, say, rejecting the generalizations
behind the awkward observations, while some reject different
generalizations, and others reject yet different ones again?

It is precisely the overall holism attributed to our theory of meaning that poses this problem. If, once more, there were a distinction
between analytic meaning-fixing generalizations and synthetic ones
(or even if there were an analogous distinction between the concept-
constituting 'hard core' assumptions in a theory and the revisable
auxiliary hypotheses) then we could suppose that the analytic (hard-
core) postulates of a theory fixed the meanings of the terms in such a
way that when new evidence appeared it was objectively indicated
which synthetic revisions were called for. But this way out is closed
once we suppose that the meanings of the terms in a theory are fixed
simultaneously by all the accepted generalizations of that theory.

Still, does a holist account of meaning really imply that all ways of
changing a theory in response to new evidence are equally acceptable?
They are all of course meaning changes. But why should some
meaning changes not be objectively preferable to others? If we take
care not to slip back into the assumption of the (equal) irrationality of
all meaning changes, and recall some of the points made in the earlier
sections of this chapter, then it is possible to explain why even from
within a holist theory of meaning some theoretical revisions are to be
preferred to others.

The crucial notion here is once more that of the relative *centrality* of generalizations. There is nothing in the holist theory of meaning to stop us deeming some generalizations more central than others in the sense explained earlier. Nor is there anything to make us reject the arguments which showed that it will always be natural and proper for scientists compelled to adjust their theories to point the finger in the first instance at the least central generalizations involved. And so here we have a straightforward enough explanation of how it is that scientists sharing a theory characteristically manage to agree about how to respond to anomalous evidence: they will share a like inclination to avoid revising central assumptions as far as possible. (A case worth special mention is where it is agreed, as it often is, that it would not be appropriate to respond to an anomaly by rejecting the observation report which produced it in the first place. The explanation will here be that the generalizations 'behind' the observation report are themselves relatively central. Thus we would not expect an observation report of the adjacency of two medium sized physical objects, like a pointer and a mark on a dial, to be lightly overturned—for the relevant assumptions about the effects of such physical objects on human eyes will play a part in implying any number of ideas about what we will see when.)

It might seem that these last remarks amount to an effective retraction of the holist theory of meaning after all. If we have to appeal to the notion of 'centrality' to explain scientists' judgments about which sentences to accept, then surely we cannot continue to maintain that the meanings of scientific terms depend simultaneously on all accepted generalizations. For what is the distinction between 'central' and other generalizations but a disguised version of the old analytic-synthetic (hard core-auxiliary hypothesis) distinction? What are the 'central' assumptions in a theory but those unrevisable principles which have the privileged role of laying down the meanings of the terms they contain?

However there are a number of reasons for resisting this line of reasoning. In the first place, the idea that central rules but not others are constitutive of the meanings of scientific terms could suggest that the relative unrevisability of central rules was somehow due to their being accorded a special kind of conventional status by competent speakers. This would be misleading. The unrevisability of central rules is not due to their having any distinctive semantic status. All the generalizations making up a theory can initially be viewed as on a par in this respect—they all function as components of the same kind in a structured theoretical representation of reality. It is perfectly consistent with this functional similarity that the different places that different generalizations fill in this structure should, as I have argued, make some rather than others stand out as the ones to go when some revision is demanded.

In any case, the notion of 'centrality' will clearly sustain only a comparative distinction, not a dichotomy. Some generalizations are more central than others, but there is no sense to the question of whether a given generalization is in itself 'central' or 'non-central'. So there is no particular point at which we might call a halt to our holism. It does not help to say only 'central' assumptions fix meanings, for all accepted generalizations can be considered to have some centrality, albeit to different degrees.

And finally, and most fundamentally, it is necessary to realize that the precept favouring the revision of less central rules is by no means categorical. It would be a mistake to hold that it was always scientifically appropriate to revise less central rules. For there can and do arise occasions where progress demands the revision of relatively central assumptions. A given line of research can always run into a degenerating phase, where the revision of 'auxiliary' hypotheses only gives rise to further empirical anomalies, and where in the end it becomes difficult even to make any more 'theoretical' advances, even to think up any more auxiliary revisions. And then, of course, it is time to turn to more central revisions, to consider adopting different lines of research. So the view that the central assumptions in a theory are distinctly meaning-constitutive and unrevisable is not only misleading and incoherent, but in the end is at variance with proper methodological practice.

When the precept to stick to less central revisions is thus suspended, then so is my explanation of what allows scientists to agree as to what new evidence shows to be wrong with previously agreed theories. But this is as it should be. Scientists do indeed differ about what is to be done when central revisions are in question. Thus, for instance, around the turn of the century there were a good number of turbulent years during which there was wide disagreement as to which of the alternatives to the classical ether theory constituted the appropriate response to the Michelson-Morley experiment and related anomalies. Even if we accept that in principle the history of science can give an indication of when to abandon old lines of research for new ones, there will always be leeway in practice for different scientists to pick different times to leave a sinking ship. And amongst those who have left there will be room for further disagreement on where to go, on which of the more or less embryonic and more or less centrally divergent alternative lines of research to opt for.

It will be appropriate to conclude this chapter with some brief remarks on the 'Duhem-Quine thesis'. According to this thesis any statement in a scientific theory may be retained 'come what may' if so desired, while correspondingly, no scientific statement is intrinsically immune from revision (cf. Quine [1951], p. 40). A number of writers, as for instance Grünbaum, have pointed out that this thesis can be construed in two ways. It can be taken merely to claim that, given any threatened statement, it will always be possible to devise *ad hoc* but logically coherent ways of saving it by reinterpreting it and the surrounding assumptions. As such the claim is generally allowed to be true but uninteresting. Alternatively the thesis can be taken as the much stronger claim that any of the logically coherent ways of responding to contrary evidence are as good as any other; that it really makes no methodological difference how recalcitrant experiences are responded to. And in this sense the thesis is generally regarded as unacceptable. (Cf. the various contributions to Harding [1976].)

The arguments of this chapter are in broad agreement with this standard analysis of the Duhem-Quine thesis. While a scientist can in principle always, so to speak, make a line of research out of any given statement and then stick to it *ad infinitum*, he is entitled to do so only when, and as long as, that line of research holds out adequate promise of progress. On the other hand the arguments of this chapter also make it clear that there is an important truth in the Duhem-Quine thesis: namely, that the distinction between inadmissible and admissible responses to contrary evidence is (*contra*, say, Grünbaum) nothing to do with an analytic-synthetic distinction, nothing to do with a distinction between those revisions which illegitimately change meanings and those which do not. It is indeed a mistake to think that all revisions are methodologically equally worthy. But it is not a matter of any semantic failings that makes some less defensible than others. And so if we understand the Duhem-Quine thesis to be referring to *semantic* rather than *methodological* possibilities then it says something which is both important and true.

Incidentally, we can now see what to say to anybody who feels that the lack of an analytic-synthetic distinction in a scientific language would merely be a matter of reprehensible vagueness; who feels that such a language would be better for a clear analytic-synthetic distinction to show exactly which bits of current theory were refuted by contrary evidence. (Cf. Dummett [1974a], p. 384.) Not only is there no methodological call for an analytic-synthetic distinction, but the presence of such a distinction would be a positive methodological disadvantage. The relative worth of alternative theoretical revisions can be decided perfectly adequately without any analytic-synthetic distinction. And so the institution of such a distinction could only serve to hamper scientific progress. It would simply place quite unnecessary constraints on future choices about how to revise theories.

5

OBJECTIVITY AND REALISM

1 OBJECTIVITY

How far has science now been shown to be *objective?*

In the last chapter I explained how scientists can, after appropriate waiting periods, reach agreed conclusions on the relative worth of different lines of research. And I argued that this explanation was quite consistent with any implicit commitment to a holist theory of meaning.

However, this is unlikely to have allayed all worries about scientific objectivity. The relativistic issues raised by Kuhn and Feyerabend did not relate only to the epistemological question of whether scientists have satisfactory standards for reaching agreed conclusions. 'Incommensurability' also raised doubts about whether scientific statements relate to an independently existing reality. In particular, it became difficult to see how competing theories could be construed as alternative attempts to represent the *same* such independent reality.

The arguments of the last chapter help little with these latter doubts. Suppose we distinguish clearly between questions of *epistemological* objectivity and questions of *semantico-ontological* objectivity. Epistemological objectivity requires the existence of standards which enable people to agree on the relative worth of statements. Semantico-ontological objectivity requires that there is an independent reality to which our statements are answerable. It is clear then that the modified Lakatosian methodology I have elaborated addresses itself directly only to the former kind of objectivity.

Indeed it could be argued that what we had in the last chapter was nothing more than a glorified *coherence* theory of truth for scientific statements. In effect what was argued was that a scientific statement is to be accepted if it is part of a coherent system in which the central judgments are consonant with our observation reports and our general understanding thereof, and in which this consonance can be maintained without simply giving up our understanding of our observation reports. So at bottom the worth of a statement depends on how it tallies with other statements. And, as with all coherence theories of truth, this seems to leave entirely out of account any idea that the worth of statements might involve their *correspondence* to an external reality.

The complaint, then, is that the arguments of the last chapter fail to come to grips with the problem of semantico-ontological objectivity. One way of reading this complaint is not so much as a criticism of the arguments of the last chapter but merely as an observation that more remains to be said. That is, it might be allowed that the last chapter is perfectly adequate on the epistemological issue, even if it says nothing positive on the semantico-ontological one: the idea would be that whether our statements answer to an independent reality is one question, while how we are to decide on the acceptability of such statements is another quite unconnected question.

Once we put it in these terms we see there is a stronger complaint to be made. How can we hope to discuss how decisions on scientific statements should be made, independently of any conception of how those statements are responsible to reality? Surely the epistemological questions must be secondary to the semantico-ontological ones. If we are interested in whether it is right or wrong to accept a given statement, then do we not need some idea of the *achievement* those statements aim at? It cannot just be a matter of showing that by following certain standards scientists can and do reach some kind of consensus. If consensus were enough in itself, then what would there be to be said against a community who reach consensus by always accepting, say, those statements which present changes of any kind as having unwanted results; or those statements which have an even number of words; or start with a 't'; or . . . ? Consensus is surely only worth having if there is some reason to suppose that the principles from which it flows actually do lead to satisfactory representations of reality. So it seems that in the end a discussion of semantico-ontological objectivity is not just something that might be added to the conclusions of the last chapter, but rather something on which any such conclusions need to rest in the first place.

2 A HOLIST REALISM

I shall apply the term scientific *realism* to non-scepticism about semantico-ontological objectivity for scientific theories, to the view that scientific theories are answerable to an independent reality.

A realist as so defined need not hold that scientific theories always, or ever, get their answers right. His initial claim is just that there is a question for them to be more or less right (or wrong) about. But given some version of such a scientific realism, given some idea of how scientific theories are answerable to reality, then we can always meaningfully raise the further question of whether certain epistemological standards for choosing theories are indeed such as to produce theories which do answer appropriately to reality. In particular, we can ask whether the epistemological standards defended in the last chapter will pass this test.

Let us first clear away one possible source of confusion. In the last chapter I did, it is true, argue that a modified Lakatosian methodology could in a sense be *justified*—namely by reference to the agreed aims of generality and anomaly-freedom, and in the light of the lessons of history. However, this was merely a matter of showing that certain procedures for conducting programmes or lines of research could be justified as an effective *means* to versions satisfying those agreed aims. What I did not say anything about was what justified those aims in the first place. Why are general and anomaly-free theories such a good thing? It was indeed observed that everybody agrees that such theories are a good thing. But the burden of the remarks in the last section was precisely that the uncontroversiality of certain standards is not by itself enough to establish epistemological objectivity in any serious sense. So what now remains to be shown is why a set of statements which is as general and anomaly-free as possible should be considered to correspond to reality better than an alternative set that scores less well in these respects.

How are we to conceive of scientific theories as being answerable to reality? There is one standard way of elaborating such a conception. But we have in effect already come across arguments which discredit it. Recall our earlier discussion of Scheffler's response to the relativism threatened by the meaning variance thesis. Scheffler's suggestion was that we should switch our attention from the senses of scientific terms to their *references*, to the independent existents they stand for. Clearly if this suggestion had worked out, there would be available a quite specific formulation of scientific realism. For if scientific terms did have determinate referential values, then, as explained earlier, we could follow Tarski in construing scientific statements to be *true* just in case the entities their constituent terms stand for are indeed related in the way those statements portray them to be. And so we could take it that scientific theories answered to reality in so far as the statements they contained were true. But Scheffler's suggestion did not work out—there were reasons to doubt that scientific terms did have determinate referential values. Indeed these reasons have now received further reinforcement. For in essence it was the apparent unavoidability of a holist theory of meanings for scientific terms that forced us to doubt whether the scientific terms had determinate references: in so far as we were unable to pick out some privileged sub-set from the totality of procedures for applying such terms, there seemed no alternative but to allow that the different entities picked out by different such procedures were all equally good candidates for what those terms stood for. So in acquiescing to a holist theory of meaning we have in effect now accepted that scientific terms do indeed lack determinate referential values. Which in turn means that we must reject the standard formulation of scientific realism in terms of the possibility of the Tarskian truth of the statements making up scientific theories. Such statements can scarcely be made true or false by the actual relationships amongst their constituent term's references, if those terms fail to refer determinately in the first place.

Fortunately the holist theory of meaning does not force us to do away with scientific realism altogether. It does stop us construing scientific theories as being made up of components each of which is possibly true or false. But it does not mean that we cannot have a more general and abstract idea of the way scientific theories are answerable to reality.

Note first that nothing that has been said need stop us supposing that there is an independent reality 'out there'. It is true that *any* such reality would not be of use to us. If reality were entirely amorphous, then it would not help with our problem—in so far as if it were infinitely malleable to all theoretical impositions there would be no sense in which one theory could correspond to it better than another. But there is also nothing in what has been said so far to stop us supposing in addition that reality possesses a definite articulated structure. And this then gives us enough to allow a conception of alternative theories as being better or worse representations of reality.

Even if we are not supposing scientific terms to refer determinately to certain of the elements out of which reality is structured, and so are not considering the specific statements making up our theories to be possibly true or false, why should we not think of our theories *as wholes* as being more or less successful attempts to picture reality? That our theories do not break down into separate components each answerable to its own bit of reality does not mean that such theories cannot face up to reality as integrated units. It is this more abstract kind of 'holist' realism that I shall defend henceforth. And the first thing I want to show is that such a realism is perfectly adequate to provide the requisite explanation of why generality and anomaly-freedom should be considered appropriate epistemological indices of the degree to which a scientific theory succeeds in representing external reality.

Take first anomaly-freedom. An anomaly arises when certain accepted observation statements contradict one or more of the generalizations comprising a theory. What the occurrence of such an anomaly shows is that reality cannot be in all respects as our theory pictures it. And we can see why this is so, quite independently of any considerations involving Tarski-type truth. In effect an anomaly represents a conflict between two sets of generalizations contained in our theory. On the one hand are the generalizations contradicted by the anomalous observations. And on the other are the generalizations 'behind' the anomalous observation reports themselves: those generalizations to which we commit ourselves by making those reports, and which specify that the *occurrence* of such reports is a reliable indicator of the matters they report on. What the emergence of the anomaly shows then is that reality cannot conform to both these sets of generalizations: for if the observed situation really had been as the latter generalizations implied our observations showed it to be, and if the entities thus imputed really did behave as the former generalizations claimed, then no anomaly could have arisen.

There is nothing in this argument that requires us to suppose that the terms involved in our theorizing refer determinately to any actual constituents of reality. And so there is no need to suppose that what is wrong with the theory is that at least one of our generalizations has been shown to report falsely on *those* entities. (The point here is not that we might not *know* which generalization was false. It is rather that we need not even suppose that our overall theory relates to reality 'tightly' enough for any specific constituent generalizations to *be* false.) Even if none of our terms succeed in hooking on to actual elements of reality, the emergence of an anomaly still shows that reality must somehow be different from the way our overall theory represents it. And so, given of course that we want theories to which reality does conform, this then suffices to explain why we should aim for theories which are as free from anomalies as possible.

Perhaps it will be in order here to digress briefly and add a few further remarks on induction and anomaly-freedom. The temporally unrestricted generality of our theories means that they make claims, not only about how the past has behaved, but also about how the future is going to behave—indeed it is of supreme importance for our practical affairs that they should do so. Yet in so far as we know our theories to be free of anomalies, it is always past anomalies whose absence we are aware of. This of course raises Hume's question—why should the success of our theories in fitting past reality be any indication of their future success? That is, why should we not anticipate the future just as well with anomaly-ridden theories as with anomaly-free ones? (Cf. Ayer [1956], p. 74.) It is no satisfactory answer here to argue that there is no real alternative to the supposition that there are laws to which all portions of reality, both past and future, conform, and so we have no alternative, if we are to make any serious attempt to anticipate the future, but to do so on the basis of the past. For even if we do grant that there is a sense in which the premiss is true, there then remains the problematic possibility that the concepts in terms of which we are naturally inclined to formulate temporally unrestricted generalizations might not be the ones in terms of which reality is law-like. (This is one of the lessons of Goodman's new problem of induction.) The fact that our concepts have seemed adequate in this respect up till now does not establish that they will continue to seem so. So we are left with the question: 'Why should the past anomaly-freedom of our theories indicate their future reliability?' It is of fundamental importance to us that it does, but there is no good explanation of why this should be so. However I shall not pursue this question any further. All extant accounts of science leave unanswered questions about induction in one form or another. So it is no special discredit to mine that I cannot explain why our future interests should be served by theories which are indicated to be accurate representations of reality only by the past absence of anomalies.

I turn now to the question of why *generality* should be considered a desirable attribute in theories. Let us first be precise about what is meant by 'generality' in this context. I originally introduced the idea of generality via Lakatos's notion of theoretical progress: theory A was more general than theory B if switching from A to B was a theoretically progressive step. Theoretical progress required the making of new predictions. And this comes with the addition of new generalizations to the currently accepted stock. So to say one theory is more general than another is to say that the former contains more generalizations than the latter. It is worth noting that generality is not here defined 'externally' by reference to the extent to which a theory makes claims about certain elements of external reality. Again, I am not supposing that the terms in our generalizations refer determinately to external existents and that generality is a matter of saying more about more of those entities. Generality is simply an 'internal' matter of one theory containing more generalizations than another.

There are, it is true, difficulties about deciding whether one theory seriously contains 'more' generalizations than another in cases where neither set of generalizations is a proper sub-set of the other, where the one theory is neither an expansion nor a contraction of the other. But, leaving these difficulties to one side, it is not hard to see why 'internal' generality should be considered a desirable characteristic of theories. For, given any plausible method of counting generalizations, a theory containing more generalizations will offer a more detailed picture of reality than one containing less. And in so far as we want a *more* rather than a *less* complete picture of reality, we will have reason to prefer the former theory to the latter. Of course there is always the danger that the extra information offered is inaccurate, that reality does not actually conform to the extra detail in the picture presented by the more general theory. But we have anomaly-freedom to cope with that. If a theory gains extra generality only by *mis*representing reality then we would expect it to be discredited by the consequent emergence of anomalies. So the relevant choice is between two *equally* anomaly-free theories of different generality. And, given this choice, it is obvious that the more general should be preferred—since (we have reason to think that) the more complete picture is as trustworthy as the less complete one, why ever should we not accept the extra information it offers?

Can we really then rest with a holist realism? Can we really conceive of scientific theories as answering to reality without thinking of their constituent generalizations as being possibly true or false?

No doubt many will remain sceptical. Surely the ideal of *truth* is basic to our understanding of science, if anything is. Perhaps a few further remarks will clarify my position here. I am not against the search for 'truth' in the general sense of believing in an independent reality and wanting our theories to represent it as it is. What I am rejecting is the specifically Tarskian notion of truth, which presupposes that specific scientific terms relate to specific elements of external reality, and accordingly conceives of representational success on the level of individual statements.

I am also not necessarily against the idea of the perfectly 'true' scientific theory which represents the general aspects of reality just as they are in every detail. There is no obvious reason in principle why we should not come by the theory which presents that picture of reality in its general aspects which corresponds exactly to how that reality is. But this is surely an unattainable ideal—surely we will never actually have that theory in practice. So any serious analysis of science needs to concentrate on those real theories which in one way or another fall short of perfection. (Would we be able to *recognize* the perfect theory if we found it? That theory would of course be entirely anomaly-free, and as general in any serious sense as any theory could be without violating anomaly-freedom. But from an epistemological point of view there would always remain, on the one hand, the inductive possibility of future anomalies, and, on the other, the heuristic possibility that more generality might not be inconsistent with anomaly-freedom after all.)

Now, the important point for my argument is that within the flawed theories that are all we will ever have it need not be a determinate matter which bit of the theory is inadequate to reality, which constituent generalizations fail to fulfil their specific obligations. It might perhaps seem that we could conceive of specific generalizations as succeeding in their specific jobs—as being individually true—just in case they would be retained in the final perfect theory, were it ever reached. But the difficulty with this suggestion is that there is no reason to suppose that the job that generalization will be doing in that final theory will be anything like the job it is doing now. Perhaps the final theory will contain a form of words which is identical with a certain form of words in current theory. But it remains to be shown that the meanings (in any sense of meaning) of those words will be the same in the two theories. And in the absence of any such demonstration there is clearly no reason why the worth of that form of words in the ultimate theory should indicate anything in particular about its worth given the way it is used now.

Take the picture metaphor seriously for a moment. Suppose our theories are blurred, distorted paintings of reality as it really is. The imperfections are not just a matter of giving certain things (people, houses, trees) the wrong characteristics. Rather they make it indeterminate in the first place which entities in the paintings are supposed to correspond to which entities in reality (who and what in the paintings are supposed to be who or what in reality) in such a way as to make it misleading to think of those paintings getting certain things wrong and other things right. But for all that they are paintings of reality all right, and, moreover, there can be good grounds for thinking one such painting is a better representation of reality than another.

Or, to put it without the metaphor, once we allow that all the constituent generalizations in a theory together contribute to fixing the contents of that theory's terms, then we need to allow that any inadequacy in the overall theory must to some extent reflect back on all those constituent generalizations. A fault in a theory does not mean just that the theory's concepts have been used to say something wrong, but rather that those concepts are inadequate to reality in the first place.

3 CONFLICTING SCIENTIFIC VIEWS

The remarks of the last section did not touch on, though they may well have drawn implicit attention to, a further possible motive for dissatisfaction with a limited, holist realism. This relates to the possibility of different scientific views *conflicting*.

It is natural to think of two scientific views being in conflict when one or more of the statements in one are *logically inconsistent* with one or more of the statements in the other. But, as explained in Chapter 3, logical inconsistency requires that the terms in the inconsistent statements have common referential values. To be logically inconsistent two statements have to start off by referring to the same entities—only then they can go on to say incompatible things about them. If two statements do not have any referential values in common then from a logical point of view it seems that they can only be divergent, but perfectly compatible, claims about different areas of reality.

But how then are we to conceive of conflicts between different scientific theories? If scientific terms do not have determinate referential values it cannot simply be a matter of logical inconsistencies as traditionally understood. Nevertheless some account clearly needs to be given. If alternative scientific views were never in any sense in conflict, then why should there ever be any necessity to choose between them? Why should we not simply uphold *all* such views, regarding each new alternative as an attempt to describe a hitherto unchartered territory?

An initial response to this problem might be to say that there is one thing which two views can unproblematically both be about—reality. Surely if different views present different pictures of reality there is a clear enough sense in which they are incompatible: two different pictures cannot both be complete and accurate representations of the same scene.

This response is cogent enough as far as it goes. The trouble is that it does not go far enough—it only allows for 'conflicts' of an extremely gross kind. A 'view' of reality in the requisite sense would comprise a conjunction of what would normally be thought of as physical, biological, psychological, economic, etc. views. Another such 'total view' would be a different such conjunction. Given that the one such 'view' did not contain the other as a proper part, they would indeed be in conflict, by virtue of their both being about everything but presenting different pictures of it. But surely we want to allow that there can be comparisons between conflicting views on a finer grid than this. Surely we want to allow that a given part of one total view can be in conflict with a specific part of another such, that scientists holding different views can disagree not only on reality *in toto* but more specifically in certain of the individual things they say about it.

I shall distinguish here between two different senses in which we need to allow that different scientific views can be in conflict, both more fine-grained than 'total' conflicts about reality. The first will relate to conflicts between different *versions* of research programmes. The second will relate to conflicts between *programmes* themselves. Having explained the difference, I shall then show how both kinds of conflict can be accommodated within a holist realism.

'Versions', you will recall, were conjunctions of generalizations specific enough to give rise to definite predictions. Now, it seems to be an essential part of science that two such versions can come into conflict by issuing specifically in *incompatible* predictions about a given experimental situation. That is, they can issue in predictions such that a particular experiment made to test those predictions can be guaranteed to decide between those two versions.

It would of course be a mistake to think that such experiments can be 'crucial' between *programmes*. Traditional exemplars of 'crucial' experiments are, say, Fresnel's interference experiment, supposed to have decided finally for the wave theory of light against the corpuscular theory, and the Michelson-Morley experiment, held to have decided similarly for special relativity over classical physics. However, work such as Lakatos's has shown that in so far as such crucial experiments are supposed to decide once and for all in favour of one programme over another, then there can be no crucial experiments. Adherents of the programme to which such an experiment presents an anomaly can always retain their programme and hope to come up with some new version which promises to deal with the anomaly progressively. (Cf. Lakatos [1970], pp. 159–77.) Nevertheless this is not to say that such experiments cannot decide at least between the *versions* which issued in the predictions in question. Even if Lorentz and Fitzgerald were able to elaborate a suggestion which promised to accommodate the Michelson-Morley experiment within the central assumptions of classical physics, the Michelson-Morley experiment was at least sufficient to decide between those versions of classical mechanics which *did* predict 'The round-trip times for the light rays will be different' and versions of special relativity which did not.

To avoid confusion here I shall speak henceforth of such experiments as *decisive* (between versions) rather than as *crucial* (between programmes). The task, then, is to explicate the possibility of decisive experiments. The difficulty of course is that the natural way of doing this is in terms of the logical inconsistency of the predictions involved, in terms of adherents of the two versions respectively affirming and denying that a certain experimental object falls within the extension of a given predicate. But this natural way is now closed, given once more that the holist theory of meaning precludes any possibility of terms from different theories being used to affirm and deny the same characteristic of the same experimental situation.

Even after the possibility of decisive experiments between versions of research programmes has been explained, there will remain a further problem about conflicts between scientific views—for the possibility of conflicts between research programmes as such will still need to be accounted for. That two versions can issue in conflicting predictions does not in itself suffice to show that the programmes of which they are versions are themselves in conflict. Consider the situation where, say, a contemporary evolutionary biologist and a contemporary geologist make different predictions about what radioactive dating will show about the age of certain uranium-bearing rocks. The biologist and the geologist will each have used their stock of accepted assumptions to make observations of the rocks and their surroundings and to reason from there to the age of the rocks. Thus the biologist's version of, say, the neo-Darwinian research programme will be in conflict with the geologist's version of the 'plate tectonics' programme, and indeed the dating procedure will provide an experiment which will decide between these versions. Yet surely we do not want to say that the two *programmes* involved are competitors. Even if they can thus manifest conflicting versions, it would be absurd to maintain that neo-Darwinism and plate tectonics are competing alternatives in the same field of research. By way of contrast consider, say, the comparison between the Ptolemaic and Copernican systems of astronomy. These two programmes also had versions which issued in incompatible predictions—for instance, early Copernicans expected to observe stellar parallax as a result of the earth's orbit of the sun, while Ptolemaists of course did not. But in this case there was the additional demand that the two programmes be decided between as such. The comparison of neo-Darwinism with plate tectonics on the other hand demands no such choice. Notwithstanding the incompatibility of the versions involved in the decisive experiment in question, it would be perfectly coherent to continue upholding both the neo-Darwinian programme and the plate tectonics programme. One cannot of course continue to stand by both the *versions* involved, and indeed the actual application of the dating procedure will show of (at least) one version that it is untenable. But there is nothing preventing scientists from simply modifying the programme which turns out to be faced with an anomaly and continuing to pursue both programmes indefinitely. So there is a clear sense in which two programmes might or might not be in conflict, over and above any conflict between specific versions thereof. (Nor is this point undermined by the observation that it often takes time to decide between two programmes which are in conflict, and that the time-lag required for the superiority of the one over the other to emerge can make it right to suspend judgement on which is to be preferred. The important thing is that on Ptolemy versus Copernicus, but not on neo-Darwinism versus plate tectonics, there is indeed a judgement to be suspended.)

But what now actually makes it the case that two programmes are in conflict? Again, the natural way of explaining this is to say that they are in conflict when the terms in which those programmes are framed refer to the same components of external reality, and that the programmes then make inconsistent claims about those things. The inconsistencies will not now just be at the level of predictions issued in, but rather in the respective sets of central assumptions defining the programmes. But the principle would be the same—the Coperni-can and Ptolemaic programmes are in conflict because their central tenets refer to the same external entities (the heavenly bodies and their motions) yet say inconsistent things thereof, while the differing central tenets of neo-Darwinism and plate tectonics merely refer to different external things and so are quite compatible. However this natural way is again closed to us, given the holist theory of meaning and the consequent doubts about scientific terms having determinate references in the first place.

4 DECISIVE EXPERIMENTS

We need to deal first with the problem of decisive experiments. We can without loss of generality conceive of such experiments on the following simplified model. One theory (version), T_1, issues in a prediction which applies some predicate, P, to some object involved in an experimental set-up. The other theory, T_2, applies a predicate $\sim Q$ to that same individual. (P and Q will of course often be the same word, but it will help to keep things clear if we assume they are not.)

The problem now is to explain what can make it the case that T_1 and T_2 are here issuing in incompatible predictions (given that we are debarred from simply saying that it happens when the 'coextension-ality' of P and Q makes it impossible for the experimental object in question to satisfy both P and $\sim Q$).

What exactly is it for the predictions in such a case to be 'incompat-ible'? In essence it is simply this: the experiment is guaranteed to fit either T_1's application of P, or T_2's application of $\sim Q$, but not both. That is, the experiment is predetermined to discredit one of the predictions and favour the other.

But once the incompatibility is described in these terms it becomes easy enough to see how we might explain it without bringing in talk of coextensive predicates. When the experiment is performed those present will report certain observations; and then certain general assumptions about the human observers, the experimental instru-ments involved, and the entities being experimented on will be used to reason from the occurrence of those observations to further conclu-sions about the characteristics of the experimental situation. So what is required for the experiment to be decisive is (1) that the assump-tions in T_1 by which one might infer the applicability of P from possible observational responses to the experiment should be the same as the assumptions in T_2 by which one might infer the applicability of

Q from such observations (abstracting of course from such typo-
graphic substitutions as Q for P), and (2) that the assumptions in T_1
for getting from such observations to $\sim P$ should similarly be the same
as those in T_2 for getting from such observations to $\sim Q$. For if this
condition is satisfied then the performance of the actual experiment
cannot fail to discredit either T_1, or T_2, but not both. The specified
similarities between T_1 and T_2 mean that the observations will *either*
lead via inferences to T_1's adherents applying P and T_2's adherents
applying Q (T_2 is in trouble) *or* to T_1's adherents applying $\sim P$ and
T_2's adherents applying $\sim Q$ (T_1 is in trouble). And this is just what
we wanted of a decisive experiment. (It might be argued that notwithstanding the requisite similarity of general assumptions the
experiment could still fail to be decisive on the grounds that the
different theoretical presuppositions of the respective adherents of T_1
and T_2 could always lead to their having different observational
experiences in the first place. This would be the kind of 'theory-
dependence' of observation emphasized by such writers as Hanson and
Kuhn. But even this possibility need not disrupt the decisiveness of
an experiment. Remember that scientific judgments about particular
situations can always, so to speak, be referred back to inferences from
the *occurrence* of certain observation reports rather than to inferences
from the *acceptance* of those reports. And if the arduous regress
mentioned in Section 2 of the last chapter is to be avoided then the
adherents of each theory will need to allow at least that their opponents made certain observation reports, even if they themselves did
not. So, provided only that T_1 and T_2 now agree relevantly in
assumptions about what is to be inferred from people with T_1's
theoretical assumptions making certain observation reports, and
about what is to be inferred from people with T_2's theoretical assumptions making certain observation reports, there is no reason why any
Hanson-Kuhn theory-dependence of observations should disrupt a
decisive experiment.)

In effect the suggestion is thus that the predictive application of P
and $\sim Q$ will allow of a decisive experiment whenever T_1 links P and
$\sim P$ to the relevant observation reports in the same way as T_2 links Q
and $\sim Q$ to those observation reports. There is no reason to doubt
that this requirement is not often satisfied. It is not of course required
that T_1 and T_2 should share all assumptions—just those which link
the relevant predicates to the relevant observations. In a sense we
might say that what is required is just that the terms should share

'observational meaning' (and we might correspondingly expect that it will be predicates traditionally considered 'observational' that allow most easily of incompatible predictions).

But if we do slip into this way of talking it needs to be emphasized that such 'observational' synonymy does not amount to any serious 'sameness of meaning' as traditionally understood, nor, correspondingly, to any guarantee of coextensionality. For there is nothing in what has been said to accord the shared assumptions linking P and Q to the relevant observation reports any superior semantic status over the other (divergent) assumptions relating to P and Q in T_1 and T_2.

In particular it is worth noting that there is nothing in the analysis of decisive experiments to stop it being perfectly appropriate for a theory discredited by such an experiment to deal with the anomaly it consequently faces by revising the 'observational meaning' of the relevant predicate, rather than by rejecting one of the assumptions that led to the prediction in the first place. Suppose it were T_2 that in the event lost out. Its adherents could always then stand by the general postulates that led to their anticipating that the situation would satisfy $\sim Q$, and change instead those assumptions that led to its responding to observations of the actual experiment with Q. Such a ploy would, if it is true, be constrained by the methodological requirements of 'progress', 'retention of central postulates', etc. But the mere fact that beforehand the theories were such that the experimental observations were predetermined to support either P and Q or $\sim P$ and $\sim Q$ does not in itself mean that they have to remain so afterwards. (Lest it be thought that an experiment that allows the defeated theory to retain its distinctive identity can scarcely be decisive, remember that what is at issue is not 'crucial experiments' between *programmes*, but merely ones which will favour one *version* over another.)

Before we do move on to the question of conflicts between research programmes themselves, there remains an obvious hiatus in my 'holist' account of decisive experiments. I have been focusing on the question of what makes the predications P and $\sim Q$ incompatible. But surely the question of their incompatibility only arises in the first place if it is taken for granted, as it has been so far, that they are predications of the same individual. It is scarcely to the point that P and $\sim Q$ are in some sense incompatible, if they get applied to different individuals in the first place. But here the holist theory of meaning presents us with another problem. If the meanings of all scientific terms depend on the theory they are used in, then what more warrant is there for supposing that a singular term (say, *a*) can denote the same experimental object as used in two different theories, than for supposing, as we are not, that predicates in different theories can stand for the same extensions?

Throughout this book I have for reasons of argument been concentrating almost entirely on predicate expressions rather than singular terms. But the problem of decisive experiments now demands at least some shift of attention. Fortunately the further issues raised will not seriously alter the analysis of decisive experiments already given.

We need first to consider Quine's view that the 'inscrutability of terms' means that there is no fact of the matter as to whether the apparatus of referring expressions in two different languages allows users of the two languages to denote the same individuals. (Cf. Quine [1960], Ch. 2. The argument for the 'inscrutability of terms' needs to be distinguished clearly from the more general arguments for the 'indeterminacy of translation', as Quine himself points out in his [1970]. Quine's more general indeterminacy thesis will be discussed in the next chapter.)

The 'inscrutability of terms' can best be illustrated in terms of radical interpretation. Suppose a radical interpreter discovers, by observation and prompting, that his aliens apply the word 'gavagai' only when a rabbit is present. How is he to know if 'gavagai' is used to denote *rabbits*, rather than, say, *rabbit-stages*, or undetatched *rabbit-parts*? (Cf. Quine [1960], p. 52.)

It might seem that the interpreter should be able to decide this by noting how 'gavagai' is used in conjunction with the aliens' numerals and their identity sign. But an awkward circularity lurks.

If . . . we take 'are the same' as translations of some construction in the jungle language, we may proceed on that basis to question our informant about sameness of gavagais from occasion to occasion and so conclude that gavagais are rabbits and not stages. But if instead we take 'are stages of the same animal' as translation of that jungle construction we will conclude from the same subsequent questioning of informants that gavagais are rabbit stages. (Quine [1960], p. 72.)

The difficulty is that we cannot decide between 'are the same' and 'are stages of the same animal' or translations of the alien construction, unless we already know what kind of things are being counted. But to know this we need to know what the denotations of at least some singular terms are—which unfortunately was just what made us look to constructions like 'are the same' in the first place.

Thus Quine argues that the possibility of 'compensatorily juggling the translation of numerical identity and associated particles' ([1960], p. 54) means that there are always alternative but equally correct ways of rendering the singular expressions of one language into another.

Now this might seem to present a problem for my analysis of decisive experiments: if there is in general no fact of the matter as to whether a singular term from one language denotes the same individual as a singular term from another, then in particular how can it be possible for singular terms from the languages of different scientific theories to denote determinately the same particular experimental object?

There is perhaps room to query whether Quine's 'compensatory juggling' can always plausibly be performed. But I shall not pursue this line of argument. For even if we do allow that there is always inscrutability in the relevant sense, it does not seriously affect the possibility of decisive experiments. The reason is that whenever alternative constructions of the singular terms in a language are imposed by 'juggling the translation of numerical identity and associated particles', the point of the juggling is precisely to ensure that *whole* sentences are always interpreted in such a way that it is rational (or at least explainable) for the aliens to assert them when they do. In particular any alternative constructions of singular terms will carry with them correspondingly alternative understandings of the predicates that get conjoined to those singular terms, in such a way that the alternativeness of the construction makes no difference to when whole assertions get made.

This is not to say that such alternative translations are identical. Rabbits are not rabbit-stages, nor are, say, interferometers interferometer-stages. But it does show why the possibility of such alternativeness is irrelevant to the analysis of decisive experiments. For suppose we analyse decisive experiments in terms of whole assertions, rather than predications alone. Then what we shall require is that the adherents of T_1 should be disposed to assert Pa in the same circumstances as dispose the adherents of T_2 to assert Qa, and similarly for $\sim Pa$ and $\sim Qa$. And to this requirement the possibility of construing a differently in T_1 and T_2 does not matter. For, as just pointed out, we then need to introduce a precisely compensating difference in the way we construe P and Q (and $\sim P$ and $\sim Q$), a difference which compensates precisely in the sense that the dispositions to make whole assertions conjoining P and Q (and ~P and ~Q) with *a* are left identical.

So we can legitimately abstract from Quinean inscrutability in connection with decisive experiments. But the real difficulty still remains to be dealt with. For one way of putting the holist theory of meaning amounts precisely to the claim that adherents of different scientific theories do *not* ever coincide in dispositions to verbal behaviour. It is all very well saying that what matters to decisive experiments is coincidence in dispositions to verbal behaviour, and that inscrutability leaves this untouched—our problem is that we cannot assume such coincidence in the first place.

In a sense, what was shown earlier was that despite overall differences in dispositions to use predicates there can be sufficient similarity across theories to ensure a kind of incompatibility between predications. What still needs to be shown is that this incompatibility can be preserved when we stop ignoring singular terms, and consider the possibility that dispositions to use singular terms can also vary across theories.

Recall the brief remarks made in Chapter 2 about the functioning of singular terms for spatio-temporal particulars such as are involved in experimental predictions. I suggested that we could take such singular terms to work by conjoining one or more predicate expressions with some kind of demonstrative convention. The idea here is that the demonstrative conventions direct attention to certain spatio-temporal regions, while the predicates then indicate what kind of, and which, individual in that region is being denoted. The problem we face now becomes clear. If the content of those predicates depends on all the generalizations in the surrounding theory, and if their extensions are consequently indeterminate, then it follows that the denotations of the related singular terms will be correspondingly indeterminate. To return to an earlier example, if it is indeterminate what class of pairs of objects satisfy the Newtonian predicate 'are congruent', then it will also be indeterminate which, if any, of the particular pairs of objects in an indicated spatio-temporal region is denoted by 'that congruent pair'. Is it supposed to be one with identical rigid rod measurements, or with the appropriate light ray transit times, or (not that there are any such) with both, or what?

This indeterminacy in the denotations of singular terms thus poses a prima-facie problem for the suggested analysis of decisive experiments. If no singular terms have determinate denotations, then *a fortiori* no two singular terms from different theories can denote the same individual, and the already suggested explication of the incompatibility of predications ceases to account for the possibility of incompatible *assertions*. However the remarks on the workings of singular terms already made suggest a natural way around this problem. Given some assertion of the form P*a*, why should we not consider the predicate or predicates 'involved' in the singular term *a* to be combined with the expression which is in the explicit predicate position? That is, why should we not consider the assertion, instead of having the originally indicated form, to be a more complicated predication, with the subject of the predication now being simply the spatio-temporal region indicated by the demonstrative conventions in *a*? Once this is done then we can apply the earlier analysis of the incompatibility of predications in general to these more complicated reparsed predications, and as before there will be many cases (such as those involving 'observational' terms) where the requisite similarities of usage across theories will ensure that these more complicated predications are incompatible in the relevant sense. And then all that remains for the incompatibility of *predictions*, for decisive experiments, is that adherents of different theories be able to advert demonstratively to the same spatio-temporal region.

The working of demonstrative conventions is perhaps itself a topic that deserves discussion. But even without going into any details it should be clear that there is nothing in the arguments of this book so far, nor in particular in those for a holist theory of meaning, to suggest that adherents of different theories cannot coincide in their use of demonstrative conventions. (I certainly do not want to imply that the suggested reparsing of the assertions involved in decisive experiments casts any light on such further matters as their ontological import or their grammatical structure. The sole purpose of this reparsing is to show what makes decisive experiments possible.)

5 PROGRAMMES IN COMPETITION

I return now to the question of what it is for two research programmes to be competing alternatives in the same field, rather than complementary views which might both deserve indefinite development. As explained, this requires something more than that the two programmes can issue in incompatible predictions, for complementary programmes can also manage that. The problem is to explain what that something more is, if it is not that the terms in the two programmes' central tenets refer to the same external entities.

At this point it will be useful to attend once more to the deductive systematization of theories. Recall that a scientist accepting a theory will not adopt all the generalizations it implies one by one: instead he will commit himself to a limited number of postulates, which then deductively generate a large number of derived generalizations. Recall also how the notion of centrality was explained by reference to the possibility of such deductive systematization: one postulate in a deductive systematization is more central than another if it is involved as a premiss in more deductions of derived generalizations. The pyramid metaphor illustrated these points. The struts of the pyramid were the postulates in a deductive systematization; the different pathways between points made possible by the pyramid were the derived generalizations of the theory; and the struts highest up the pyramid, through which many such pathways had to go, were the most central postulates.

Suppose now we were to subtract from our 'total' current theory, from our total current stock of assumptions, all those assumptions above a central level of generality. That is, suppose all the derived generalizations in our current pyramid were filled in, and then the top of that pyramid were lopped off. By definition, what would be left would be capable of only a diminished degree of deductive systematization. But nevertheless almost all the original set of derived generalizations would remain. Now we can consider alternative research programmes (or, more usefully, alternative lines of research) as suggestions for possible ways of restoring high-level postulates once more to lend deductive systematization to the remaining set of derived generalizations. (Recall that lines of research were defined in terms of the acceptance of sets of postulates above a given level of centrality.) And then, given any two such suggested lines of research, we can consider them to be *competing* alternatives, rather than *complementary* ones, if they are intended to systematize the same part of the remaining set of derived generalizations, rather than different parts. A slightly modified metaphor will make this idea clearer. Consider our 'total' theory not as a single pyramid but as an area of adjacent pyramids, each pyramid deductively systematizing a sub-set of our total stock of derived generalizations. Then consider all those pyramid tops to be lopped off. The suggestion is that two alternative pyramid tops are competing alternatives if they are both designed to go on the same topless pyramid, whereas they will merely be complementary if they are designed to cap different ones.

The obvious drawback to this explanation of what makes for competition between programmes is that it presupposes that what generalizations are accepted below a given level of centrality is independent of what generalizations are accepted above that level. But this presupposition is untenable. The main point of analysing scientific development in terms of research programmes in the first place was that the lower level assumptions which will substantiate a given choice of central assumptions are not in general available prior to the making of that choice. Somebody adopting a given research programme is committed to *constructing* a set of auxiliary hypotheses which will support that programme's hard core. And different hard cores, even those from competing programmes, will demand the construction of different structures of auxiliary hypotheses.

Nevertheless the essential features of the suggested solution can withstand this objection. The crucial point is that the scientists pursuing a given programme will by no means have complete freedom as to how to construct a supporting structure for their hard core. For a start they will be obliged to provide a supporting structure which makes the eventually resulting pyramid come out both general and anomaly free. And this implies in particular that that pyramid should relate a wide range of possible observations to each other, and do so in a way which tallies with the actual observations that have been made. (As noted before, neither observations nor things observed comprise everything; but they are something.) So it is a constraint on the eventual substantiation of a given programme that it be filled out in a way that as far as possible accommodates the patterns displayed by extant observations. This constraint does not in itself guarantee that competing programmes will at any particular level coincide in the generalizations they contain. For observations themselves can always be reinterpreted, and previously accepted observation reports can be rejected. But, as was shown in Section 2 of the previous chapter, there are in the end significant restrictions on the possibility of such reinterpretations. So in practice we will find that with competing programmes there will be some range of observation reports, even if they are only observation reports about observation reports, where the adherents of the competing programmes will coincide in their observational application of terms; and at this point, wherever it comes, the two programmes will be aiming to systematize the same set of generalizations framed in those terms. Whereas with complementary programmes there will either be no such common observational terms, or, if there are any such, the two programmes will be aiming to accommodate different kinds of patterns manifested by their use. Thus the Ptolemaic and Copernican programmes can each be considered to be after systematizations which would include the same pattern of reported regularities about the 'positions' of 'heavenly bodies'. On the other hand, even though the geologists and evolutionary biologists have some common observational usages ('geiger-count of n per min.'), the generalizations involving such terms which their programmes aim to accommodate are different.

It is important to be clear about what this argument amounts to. The idea is not that there is some 'observational' level of language where the terms of competing programmes will share meanings. I am, it is true, arguing that at some level there is an identical structure of generalizations which both programmes are in prospect going to contain. But this is not to say that those generalizations will have the same meanings in both programmes. When we lop off the tops of our competing programmes to lay bare the identical structures at the bottom, we in a sense change the meanings of the components of those bottom structures. That is, the (identical) meanings that those components would have as parts of that common bottom structure in isolation are different from the actual meanings they do have as parts of their total respective programmes. What would be left if the tops were lopped off would be the common 'observational' meanings of the terms used at that lower level, the common way they are used in relatively direct response to observational inputs. It is precisely because those terms share such 'observational' meanings that the two programmes are in prospect constrained by actual observations made to accommodate themselves to the same structure of foundational generalizations. But as was made clear in the last section, to say that two terms share such 'observational' meanings is not to say that they share meanings in any substantial or traditional sense.

It is worth emphasizing that the foundational coincidence of competing research programmes is only a coincidence *in prospect*, not one that need be present while the programmes are competing. For one thing, there is the original observation that research programmes do not start as completely structured pyramids; the whole point of thinking in terms of research programmes is that the supporting structure for the uppermost hard core is constructed only after the hard core has been adopted. In addition, it needs to be allowed that when competing programmes start developing auxiliary hypotheses they can well conflict in the generalizations they hypothesize at the lower levels. But what then happens is that such conflicts give rise to decisive experiments which decide for the current version of one programme over that of the other. And so, provided we keep going down to levels at which 'observational' meanings do remain shared, there will be a tendency in the limit for two competing programmes to end up with the same structure of bottom generalizations.

As we saw, the possibility of decisive experiments did not in itself ensure that two programmes were in competition. But if two programmes are such that decisive experiments constrain them in the limit to accommodate themselves to the same foundational structure, then there will in the end be a need to choose between them. Non-competing programmes can happily be developed indefinitely in different directions, for, despite isolated decisive experiments, their eventual destinies are unconnected. But with competing programmes a choice must eventually be made, for it will simply not be possible for them to remain different and yet fit equally well on top of the same base.

This now accounts for competition between programmes without any need to suppose that the central (or any other) tenets of competing programmes are framed in co-referential terms. The account is entirely 'internal' in the sense that what ties competing programmes to each other is a certain prospective similarity in their structure of generalizations, not any putative common relation to certain entities in the external world. I am not of course denying that the central tenets of such research programmes are aiming to characterize real existents, a grasp of which will allow a deductively systematic understanding of the matters reported in the less central tenets. But we can allow this and yet doubt that any actual programmes succeed in determinately talking about such existents, and so *a fortiori* doubt that any different programmes manage to talk about the same such existents.

6 PARTIAL REFERENCE

The conventional notions of truth, falsity, and logical inconsistency presuppose determinate referential relations between scientific terms and the actual elements of external reality. I have argued that even if there are no such relations we can still retain enough realism to justify epistemological standards, and to explain in what senses different scientific views can be in conflict.

But by this point some readers will no doubt be becoming suspicious of the initial admission which has driven me to these argumentative lengths. Perhaps we can construct substitutes for truth, falsity, logical inconsistency, etc., even without assuming that scientific terms refer. But, in comparison with the natural way of conceiving of these matters, these substitutes do seem somewhat ersatz, do seem contrived replicas of the real thing. So perhaps it is time to review the arguments which led to doubts about the referential properties of scientific terms in the first place.

I shall do this indirectly, by discussing two recent suggestions which in different ways promise to overcome the relevant difficulties and reinstate a referential understanding of scientific terms. In the rest of this section I shall examine Hartry Field's notion of 'partial reference'. In the next section I shall turn to Saul Kripke's and Hilary Putnam's arguments for adopting a 'causal theory of reference' for certain scientific terms. However neither of these suggestions will fulfil their promise, and we shall have to rest content with the holist realism developed so far.

In his 'Theory Change and the Indeterminacy of Reference' ([1973]) Field accepts the initial arguments for doubting that scientific terms have determinate referential values. That is, he allows that certain postulates involving a given term in a given theory will characteristically indicate one referential value for that term, while other sets of postulates will indicate that the term has a different referential value; and he allows that in the absence of anything to make one such indication preferable to others the term cannot be held to refer determinately to any one of the entities in question. But he suggests that the term can nevertheless be conceived of as *partially referring* to *each* of those entities.

Field takes as his primary example the Newtonian quantity term 'mass'. He points out that from the point of view of relativity theory there are two equally plausible candidates for the referential value of this term: *relativistic mass*, which is equal to total energy/c^2 (and so increases with velocity), and *proper mass*, which is equal to non-kinetic energy/c^2 (and so is velocity-invariant). If we concentrate on such Newtonian principles as 'momentum equals mass times velocity', then we will be inclined to take Newtonians to have been referring to relativistic mass, for what relativity theory tells us is that momentum is given by relativistic mass times velocity. On the other hand, if we attend to such Newtonian assumptions as 'mass is invariant with respect to change of reference frame', then we will be inclined to conclude that Newtonians must have been talking about proper mass, for it is this, and not relativistic mass, that relativity theory indicates to be so invariant. (Field [1973], pp. 463–9.) So Field concludes that it would be wrong to say that the Newtonian term 'mass' definitely ('fully') denoted relativistic mass, or that it definitely denoted proper mass. But, he urges, this still allows us to conceive of Newtonian 'mass' as partially denoting each of these quantities (p. 476).

The point of introducing partial reference in the face of the indeterminacy of full reference is that it allows Field to suggest an alternative to the now unviable traditional Tarskian explication of truth and falsity for scientific statements. For any scientific statement, consider all the different hypothetical ways full referential values could be assigned to the constituent terms, subject to the constraint that any such hypothetical full referent is really one of the partial referents of the relevant term. Field's suggestion is that a scientific statement is *true* (*false*) if and only if the old Tarskian explication *would* have made it true (false) for every such hypothetical way of construing its terms' partial referents as full referents. A statement which the Tarskian explication would have made true for some such constructions and false for others lacks a determinate truth value. The idea is thus that a statement containing partially referring terms is definitely true (false) only if it would 'come out' true (false) whichever amongst the partial referents we mis-took those terms to be fully referring to, while statements for which no such univocal answer is forthcoming are neither true nor false.

So Field hopes to use the notion of partial reference to reinstate our natural inclination to think of the specific statements making up a scientific theory as possibly true or false. Even if the simple Tarskian analysis is still ruled out, the partial reference approach does promise to take us some way beyond the conception that our theories answer to reality only as integrated wholes, towards the idea that scientific terms hook on to certain external entities in such a way that the correctness of a specific statement involving such terms depends on whether those entities are related in reality as the statement portrays them to be. (Though Field does not discuss the matter explicitly, the partial reference approach would obviously also allow a kind of reconstruction of inter-theoretic inconsistencies and other logical relations.)

At first sight Field's approach might seem vulnerable to the objection that attributions of partial reference are themselves theory-dependent. Thus it could be allowed that from the point of view of relativity theory the Newtonian term 'mass' seems to refer partially to both relativistic mass and proper mass. But do not things come out quite differently from, say, the point of view of Newtonian physics itself? Presumably a Newtonian would take his term 'mass' to refer quite fully to (Newtonian) mass (just as Einsteinians take their term 'relativistic mass' to refer fully to relativistic mass and their 'proper mass' to refer fully to proper mass). And presumably a Newtonian would conversely take both the Einsteinian terms 'relativistic mass' and 'proper mass' to refer partially to his (Newtonian) mass (since Einsteinians on the one hand accept assumptions such as 'momentum equals relativistic mass times velocity', and on the other such assumptions as 'proper mass is velocity-invariant') and partially to nothing (Einsteinians hold in addition that 'relativistic mass increases with velocity' and that 'proper mass does not in general equal momentum divided by velocity'). But then, if attributions of partial reference depend in this way on the theory we start from, surely there is no serious hope of using the notion of such partial reference to reconstruct any traditional conception of a scientific statement being true or false according as it correctly pictures some impartial external reality.

As it stands this objection to Field's theory would be quite misconceived. The notion of partial reference is not supposed to explicate some relation between the terms of different scientific theories (such as Newtonian physics and relativity theory). It is intended rather to explain how the terms of a given scientific theory (for example, Newtonian physics) might relate to the constituents of reality. The ultimate aim is to give some account of how the statements of, say, Newtonian physics function as representations of external reality. To this end we want to consider how, if at all, the terms of Newtonian physics might 'hook on to' certain constituents of that reality. When we look at it from this perspective it should be clear that relativity theory as such need not come into the matter at all.

However the objection in question does raise a number of issues that require clarification. When Scheffler's referential response to the paradox of meaning variance was first introduced in Chapter 3, it might have seemed to some readers essentially question-begging. Scheffler's idea was that we would be able to dissolve problems about the objectivity of scientific theories by considering the external references of scientific terms. But did this not illegitimately presuppose some kind of theory-independent access to the external world by means of which those referential values could be ascertained in the first place?

The soundness of this accusation depends on whether the worries that the appeal to references is supposed to allay are worries about epistemological objectivity or about semantico-ontological objectivity. In so far as it is epistemological objectivity that is at issue then the appeal to references is indeed question-begging. If we are concerned about scientists coming to agree about which statements are to be accepted, then if anything it will be what they *take* to be the references of (their own and others') terms that will matter. For (in so far as we want to think of scientists as working in this way) it will be their *judgements* of what scientific terms refer to that will influence their decisions as to which statements involving those terms are true. But then there is the problem that scientists accepting different theories will come to different conclusions about what (their and others') scientific terms refer to: their different theories will naturally mean that they have different views about what entities populate the world and what the characteristics of those entities are, and so they will have different views of what is available for scientific terms to refer to in the first place. There is indeed no theory-independent way of ascertaining the references of scientific terms. And so it is true that thinking in terms of reference will not get us any closer to understanding what makes scientific agreement possible. However, if we consider the appeal to references as a response to worries about semantico-ontological objectivity these considerations lose immediate relevance. For then it is not what scientists theory-dependently *take* to be the references of scientific terms that matters, but rather what those references *are*.

Although I did not pause in Chapter 3 to distinguish between the different worries about objectivity that the appeal to references might be intended to answer (in his [1967] Scheffler seems to suggest it will help with both), my initial argument against the appeal told against both ways of construing it. For my argument was not simply that from within different theories we will get different ideas about what, if anything, various scientific predicates refer to. It was rather that from an 'external' standpoint there is no fact of the matter about what they do refer to: the relation between scientific predicates and reality is simply not such as to univocally select determinate classes of individuals as their extensions. And it is from just this point of view that we should understand Field's notion of partial reference. His argument is not that once it is allowed that scientific terms can partially refer to different entities it will somehow become possible for scientists from different camps to *agree* about what they do so partially refer to. The claim is only that from an 'external' standpoint there are entities to which they *do* partially refer.

Even so, it might still seem puzzling that Field, and I myself, assume relativistic physics in elaborating the examples which are supposed to show that classical terms ('mass', 'are congruent') lack determinate full references, but might yet have partial ones. If relativity theory is not supposed to come into the matter, what is it doing here? But, once again, it is important to remember we are not comparing classical and relativity theory with a view to deciding which is to be accepted. In the present context we can assume that this epistemological question has already been answered in favour of relativity theory. We are trying to work out how scientific terms relate to reality, and as particular examples we are considering certain classical terms. In conducting this investigation we will naturally and quite sensibly take reality to be as it is said to be by the theory we (with good epistemological reason) currently accept. If we are trying to identify relationships between linguistic expressions and parts of the non-linguistic world, then what alternative do we have but to work with the best-going theory of the latter? ('We should use *someone else's* conceptual system?', Putnam [1975], p. 192.)

However there is one thing that we would *not* be entitled to take for granted in the present context of discussion: namely, that the terms in our currently accepted theories refer fully to external existents. We indeed have no alternative but to assume our current theories in investigating what the terms of past theories might have referred to. But what the results of such investigations in general seem to indicate is that those terms did not have determinate full referents. So, if past form is anything to go by, the terms of our current theories do not have determinate full references either. This might seem paradoxical. How can we assume current scientific theory and simultaneously deny that its terms refer to anything in reality? But there is no reason why we should not learn to perform this balancing act. We have already this century learned to live with the idea that in all probability much of currently accepted scientific theory is actually false, without this stopping us accepting it. As before, what are we supposed to accept instead? But if we can accept theories while allowing that they are likely to be false, why should we not also manage to accept them while allowing moreover that their terms probably lack determinate referents? Of course in accepting a theory we in a sense presuppose that its terms have full referents. But, even more obviously, in accepting a theory we presuppose its truth. I can see no reason why such acceptings should be less able to withstand doubts about the former kind of presupposition than they can withstand doubts about the latter.

This last observation is relevant to the assessment of Field's approach. I shall argue that in the end Field fails in his intention of showing that (many) scientific statements are determinately true or false. The difficulty will be that, even if we allow the idea of partial references, there is no reason to suppose that any statements will have anything other than indeterminate truth values, that is, be true according to some partial references and false according to others.

Field admits that the terms of our current theories are likely to lack full referents ([1973], p. 480). But he does not seem to take this admission to heart. For when, at the beginning of his article, he argues for the necessity of some such analysis as his, it is on the grounds that:

> . . . we'll see that there are sentences with perfectly determinate truth values which contain referentially indeterminate names and predicates, so that it makes perfectly good sense to ask whether the sentence is true or false even though it doesn't make sense to ask what the name really denotes or what the real extension of the predicate is. (p. 463.)

But the cases we then come across are (of course) only examples where current theories indicate that the terms in certain sentences of past theories had partial references which would (*modulo* Field's analysis) make those sentences come out definitely true. The trouble is that this by no means shows 'that there are sentences with perfectly determinate truth values'. For if the terms of current theory do not fully refer, as they probably do not, then the actual partial referents of the past terms in question will be different from what current theory indicates, and despite current indications the past sentences may not be determinately true (even *modulo* Field's analysis) after all. If we already had reason to accept the partial reference approach to truth, and so wanted to come to some specific conclusion about whether some past sentence was true, then, as before, we would have no alternative but to let current theory tell us what is available for the terms in that past sentence partially to refer to. But, as Field explicitly indicates, there is a prior and more general question of whether 'it makes . . . good sense to ask whether the sentence is true or false' in the first place. His argument is that since we have examples showing it does so make good sense we consequently need something along the lines of his partial reference approach. But in this context of argument it is no good having examples where it is current theory which suggests certain past sentences are definitely true. For there remain good reasons for doubting that those sentences *are* definitely true. And such doubts then remove the initial rationale for adopting the partial reference approach: if in actuality no sentences are definitely true or false then we surely do not want to adopt an analysis which will lead us to impute such features to them.

An independent line of reasoning can be used to reinforce the claim that actual partial referents of scientific terms will always be such as to prevent sentences containing them being definitely true or definitely false. What exactly is supposed to make some entity a partial referent of some term? We can see the matter as follows. If we take all the surrounding tenets in the theory in which the term was used, then we will find that no determinate entity is picked out as its referent. But if we 'select' some of those tenets, and 'drop' the others, then there will be a univocally indicated referent. The partiality of the reference then arises from the fact that different such 'selections' will univocally indicate different such referents. But the question that now needs asking is: exactly which selections are admissible?

There is some suggestion in Field's article that what matters are alternative ways of selecting from the *central* tenets in a theory (pp. 467–8). But, as we have already seen, centrality is not a qualitative matter: even if we allow, as we must, that some tenets are more central than others, there is no cut-off point at which centrality stops. As before, appeals to centrality fail to isolate semantically privileged sub-parts of theoretical structures.

It seems inevitable that if some sub-selections of a term's surrounding tenets are to be allowed to pick out a partial referent for it, then they are all going to have to be allowed to do so. If there is no good basis for discrimination, then why ever should any amongst the possible sub-selections be disqualified? But this suggests then that for any given term there will be a wide range of partial referents, a far wider range than Field himself seems to have in mind. And it seems correspondingly unlikely that any sentence will be such that all the possible combinations of ways of assigning partial referents to its terms will alike make it 'come out' true (or alike make it 'come out' false).

There is also the possibility that many sub-selections for a given term will indicate it refers to nothing. (Consider the sub-selection for 'congruent': 'congruent bodies give identical rigid rod measurements' *and* 'congruent bodies with different motions give different light transit times'.) We certainly unreflectively allow that certain terms ('phlogiston', 'telepathic') fully refer to nothing. So why should we not allow that they can so partially refer? But then it seems that many sentences will lack definite truth values for no other reason than that on some assignments of partial references some of their terms will be non-referring. (Field avoids the question of exactly what the presence of such partially non-referring terms would mean for the truth value of the containing sentences, by arguing that the interesting cases of past scientific terms could not possibly have referred to nothing. But his argument is in the end simply that if we did suppose this then we would, on any account of the import of non-referring terms, be precluded from counting as true some past sentences which clearly were determinately true ([1973], pp. 470–2). Since I have offered independent arguments for doubting whether any scientific sentences are determinately true, even given appearances and the notion of partial reference, this argument ceases to carry any weight.)

So, all in all, the notion of partial reference fails to reinstate our intuitive belief that individual scientific statements can be true or false. The proffered examples of such statements beg the question in favour of the partial reference approach. And when we turn to the question of what actually fixes partial references we find further reason for doubting that partial reference helps. Note that the argument is not that there is anything incoherent or initially misguided about the notion of partial reference or the associated truth definition. It is just that the upshot of that definition is that in all likelihood no scientific statements are true and none are false, but all alike of indeterminate truth value. But this does mean that Field's truth definition is no real improvement on Tarski's as an account of semantico-ontological objectivity. All scientific statements, and the theories they compose, will bear exactly the same relation to external reality (indeterminate truth value), and we will have been offered no account of what it is for some to answer to reality better than others.

7 THE CAUSAL THEORY OF REFERENCE

All the approaches to meaning considered so far have in one way or another presented reference as somehow dependent on sense. The reference (partial references) of a word is (are) the *external* existent(s), if any, it stands for. The sense of a word is a *mentally* graspable something associated with the word (an image, a sensory experience, a concept, a set of analytic postulates, a theory). And reference has depended on sense in the obvious way: the reference of a word is picked out as that individual or class, if any, which conforms to the ideas in the word's sense.

The causal theory of reference represents a radical departure from this way of thinking. It suggests that the references of at least some words are fixed quite independently of any mental senses we may attach to those words. As Putnam has put it, ' "meanings" just ain't in the *head*!' (Putnam [1973a], p. 704).

Let us start with the kind of example with which Kripke initially elaborated the causal theory in detail, namely, proper names for people, like 'Bertrand Russell', 'Winston Churchill', etc. (Kripke [1972]). The most popular pre-Kripkean view would perhaps have been that such a proper name referred to the person it did by virtue of the user associating it with some description or family of descriptions: the person denoted would then be whoever fitted (most of) those associated description(s). Kripke however holds that we can *refer* to someone by a proper name even if we know no descriptive facts whatsoever about that person, and indeed even if we have a number of positively false descriptive beliefs about him. This is because he takes the reference of a proper name to be fixed as follows. He supposes that there was an original baptism at which the person was picked out and dubbed with his or her name. Those present at the baptism will then use the name in communication with others, and they in turn will pick it up and pass it similarly on to further speakers, ... So when any member of the community currently uses the name, his so doing will be the terminal point of a 'causal' chain starting in the original baptism, and continuing, via other speakers' uses of the name, up to his present use. And it is simply *this* that, according to Kripke, makes the name in this later speaker's mouth refer to the person originally dubbed with it, simply that the word's being there is a terminal point of such a causal chain going back to that person. Note that the later speaker need not know any facts at all about the person he is naming, need have no way of identifying him or her; indeed he may have only false beliefs about that person and consequently may be inclined to *mis*identify him or her. Yet that person would still *be* the person being referred to simply by virtue of standing at the beginning of the causal chain terminating in the later speaker's use of the name.

This account of proper names for people fits our intuitions far better than the traditional description account. For example, if someone thought that 'Bertrand Russell' names the philosopher who wrote the *Philosophical Investigations*, and had no other thoughts about who that name stood for, we would not therefore be inclined to allow that when he said 'Bertrand Russell wrote the *Philosophical Investigations*' he said something true. But so we should, if we thought that the person he referred to by 'Bertrand Russell' was the person who best fitted the sense he associated with that name.

However, let us leave proper names for the moment and consider how the causal theory of reference might apply to scientific terms. Kripke extends his causal account of reference to those scientific terms which stand for natural kinds, such as names of material stuffs ('gold', 'lead'), biological species ('tigers', 'elms'), and physical quantities ('heat', 'electric current'). Here too the story is that there was an original baptism, involving some sample of the stuff (member of the species, manifestation of the quantity), where the stuff (species, quantity) was dubbed with its name. And again the idea is that any later uses which are causally descended from that original dubbing succeed, precisely by virtue of that descent, in referring to that stuff (species, quantity).

Now, it should be clear enough why this is supposed to help with the problems posed by a holist theory of meaning. (Kripke himself has not applied the causal theory to such problems. But see Putnam [1973b].) On the causal theory, what a scientific term refers to need have nothing to do with what the users of the term currently think, nothing to do with what theoretical assumptions involving that term they presently adopt. The references of their terms are not picked out as those things that satisfy those assumptions, but simply as those things which were involved in the baptismal occasions when the terms were first introduced. So the problem of deciding which bit of theory fixes reference disappears (no bit does), and therewith our reasons for doubting that scientific terms have determinate referential values. Indeed there seems no remaining reason for doubting that scientific terms from different theories can have the same referential values: all that is required is that the use of those terms stems from the same causal origin for both theories' adherents.

Unfortunately there are various gaps in the causal theory of reference as presented. And once these are filled in it will become apparent that it cannot fulfil the promise of an easy escape from our earlier problems. The first thing to note is that even if an initial baptism need not tie any specific descriptive criteria to the term that is being introduced, what does need to be aired is what *type* of thing is being named. After arriving in a new country a group of travellers come across a native. Their leader says 'Let's call this one "Hamlet" '. Unless it is somehow specified what type of 'one' is here being dubbed, there would be nothing to ensure that it is a *person* that is being named, rather than a *race* (e.g. the Nordic race), or a *species* (homo sapiens), or a *skin-colour* (greyish-pink). Of course features of the context could somehow make this clear without any overt specification. But the point remains that it does somehow need to be made clear. It is one thing to allow that the reference-fixing procedure need not involve any specification of descriptions which will serve to distinguish Hamlet from other people (assuming that is indeed the type of thing being named). But what the example shows is that, even so, it does at least need to be specified that it is a person being named in the first place.

We can take the concept of a person to be that of a certain type of animate entity whose successive physical stages are continuous through space and time. The point just made is that it is only in so far as this concept is implicit in the baptismal ceremony that a *person* gets named. And putting it in these terms it is easy enough to see why this should be so. For what is present at the ceremony is a *stage*, a 'time-slice', of a person. To get from this to what is being named we need some principle to specify what sort of larger entity this person-stage is supposed to be an exemplification of. It is this that is given by the concept of a person. Different concepts (that of a race, a species, a skin-colour) will indicate different ways of 'generalizing' or 'extrapolating' from the same person-stage to a larger entity it belongs to. And so if a different such concept were involved in the baptism we would have a different entity being named.

Now essentially the same point can be made about natural kind terms in science. It is all very well saying that the reference of 'gold' is fixed as whatever was originally dubbed with the name, quite independently of what anybody might believe that referent to be like. But this claim is quite unspecific without some further information of what type of thing was there being named. What needs to be specified is that it was a material stuff that was being baptized, rather than, say, a colour, or a statue, or a denomination of coin. (I shall restrict myself to stuff terms in the following remarks. But essentially the same arguments would apply to species and physical quantity terms.)

This now raises the question of the nature of our concept of material stuff. What is it for different samples to be samples of the same stuff? I explained what it is for different stages to be stages of the same person, and how the presence of this concept at the introduction of a name could ensure that it was a person that got baptized. But what does the same job for stuffs?

Kripke and Putnam suggest that the requisite similarity is being similarly composed out of the basic physical constituents of the universe. Thus, for example, they suggest that what makes all samples of gold examples of the same stuff is simply that they are composed similarly with respect to their constituent atoms, electrons, etc.

But does not this now stop the causal theory of reference from being of help with our difficulties about scientific terms? For what happens when different groups of scientists have different ideas about the basic constitution of some material stuff? If what stuff terms stand for depends on such ideas, then surely we will once more be without any guarantee that the same things will answer to the different assumptions involving a stuff term in different theories. Indeed what assurance will we have that there will be anything in reality answering determinately to any of their various sets of assumptions? To take an example, in the early years of this century there were differences of opinion as to whether the term 'lead' applied only to the stable matter with atomic weight 206, 207, or 208; or whether it applied in addition to the radioactive substances with atomic weights 210, 211, 212, or 214. The disputants here seem to have had different ideas about what is required for something to be the same stuff as the samples originally dubbed 'lead' (which presumably were predominantly composed of lead 206, lead 207, and lead 208). But then surely the Kripkean story implies that if they were using the term 'lead' to refer to anything, they were using it to refer to different categories of matter.

However an advocate of the causal approach could with some justice complain that this line of argument misses the point. It is not what scientists *take* to make something the same stuff as the original sample that fixes the reference of a term like 'lead'. To suppose this is to slip unreasonably back into the old way of thinking which had reference depending on the sense users attach mentally to a term. It is rather what it *is* to be the same material stuff as the original sample that fixes the reference of 'lead'. That different groups of scientists might at different times have different ideas about the properties of lead is neither here nor there. All it shows is that those groups (bar perhaps one) are wrong about what lead is like, and consequently no doubt in danger of misidentifying it. (Remember once more that an account of how scientific terms refer is designed to deal with questions about semantico-ontological rather than epistemological objectivity. How scientists are best to choose assumptions involving the term 'lead', say, and how they are best to identify samples that satisfy it, are epistemological questions. We are concerned here with prior questions of what those assumptions are *about*, or what *does* satisfy 'lead'.)

This defence of the causal approach would be a valid retort to the objection as originally presented. However the suggested objection can be made to bite at a deeper level. For scientists do not differ on just such specific questions as what physical constitution is characteristic of lead, as opposed to other stuffs. They can also differ in their general concept of a material stuff itself, on the general question of what it is for any two bits of stuff to have the same physical constitution. The Greeks took the basic physical entities to be the four elements of fire, earth, air, and water. Dalton took them to be the indivisible atoms of the various chemical elements. During this century we have come to recognize an expanding range of sub-atomic particles. Differences on this level naturally give rise to differences in notions of what it is for two bits of matter to be the same stuff. And such differences do undermine the causal account of natural kind terms.

For, as I argued earlier, even if specific ideas about what differentiates lead, say, from other substances play no part in fixing the reference of 'lead', the general notion of what it is to be a given substance does play a crucial role. It is only in virtue of some such general notion that one way rather than another of 'generalizing' from the original baptismal sample is laid down. And so when different groups have different such notions they will extrapolate from such original samples in different ways: for each group the larger entity that was exemplified in the original sample and named in the baptism will come out differently. And so it is quite possible for theoretical differences about the principles which identify kinds of material stuff to result in different groups of scientists using a term like, say, 'lead' to refer to different things. Indeed it is possible to see the early twentieth-century dispute about 'lead' in just this light. The development of theories about the sub-atomic structure of the atom raised the question of what exactly was sub-atomically required for two bits of matter to be the same substance: and in particular there was the question of whether it required atoms with exactly identical nuclei or whether similarity merely in respect of the nuclei's positive charges would suffice. The latter answer would indicate that the appropriate generalization from the 'original sample' of lead was to the category comprising all bits of stuff with the same chemical properties as that sample; while the former answer would demand in addition that the bits of stuff in the category extrapolated to also shared atomic weights with the original sample. So here we would have an example of a general difference on stuff identity principles leading to a difference in the reference of a specific stuff term.

At this level of argument it will no longer do for an advocate of the causal approach to complain that it does not matter what scientists *take* to make two things the 'same stuff', that the important thing is what it is for them to *be* the 'same stuff'. For we are now, as we were not before, left without any further story about what such questions are supposed to be about in the first place. The point of the causal approach, from our point of view, was that it explained how the reference of 'lead' could be fixed independently of assumptions involving the term 'lead': the original sample plus the concept of 'same stuff' did the trick. But nothing has been said about what fixes the references of 'same stuff' if it is not assumptions involving that term. Consider the issue behind the 'lead' debate: is it right to think of classes of things with chemically identical atoms as the 'same stuffs', or only classes of things with mechanically identical atoms? The problem is not just one of how to decide this question. It is rather that we have been given no account of what it is about, no real reason to think that those who gave different answers were really talking about the same thing.

No doubt in the final, ideal account of reality there will be various principles for using the words 'same stuff', 'same substance', 'same basic physical constitution', etc. But as observed in a similar context earlier, it cannot without further argument be assumed that our present use of such terms is such as to make them determinately answerable to the facts they will stand for in the final theory. Is it clear that what the Greeks would have meant by the 'same substance' would have made coal, graphite, and diamonds all the 'same substance'? In so far as their ideas about composition in terms of the four elements have any grip, why should these not be counted as 'different substances'? After all, their different crystalline structure does mean that coal burns, graphite rubs off, ... Again, is it obvious that the nineteenth-century notion of 'same substance' was such as to make the two isotopes of chlorine, or the various isotopes of lead, all the 'same substance'? If anything, nineteenth-century chemists required identity of atoms for similarity of 'substance'. But if, as these examples show, we cannot presuppose that past uses of 'same substance' answered to the same things as our present use of this term, then why should we suppose that either past or present uses answer to the same things as the final one will?

So the possibility of variation in assumptions about stuff identity principles means that the causal theory of reference cannot assure us that scientific terms refer to the same things as used in different theories. But it might nevertheless seem to suggest a way of getting some distance beyond the holist theory of meaning. For can we not construe such stuff identity principles as at least playing a privileged role in fixing the references of scientific terms? This would scarcely be something intended by the original proponents of the causal account. For it clearly concedes the point that reference depends in part on what assumptions are accepted. But still, if we did allow that kind identity assumptions played a special role in fixing references, then the original 'holist' argument for doubting that scientific terms can have determinate references would fall away, and there would be the possibility of common references whenever such assumptions are shared.

However, not even this suggestion stands up to examination. To see why, we need to consider further what is involved in the adoption of identity principles for some type of kind, such as material stuffs, or the biological species, or the physical quantities. (I shall revert now to a general discussion of the different types of natural kinds.)

A given natural kind consists of sub-entities (bits of lead, tigers, amounts of heat) each of which share a cluster of properties. In theorizing about a natural kind we will correspondingly have a number of generalizations stating the various properties common to all the examples of that kind. These properties will include both 'manifest' (common-sense, observable, pre-theoretical) properties such as colour, shape, tactile impression, etc., and underlying 'structural' properties, like atomic constitution (stuffs), or genetic make-up (species), or basic physical activity involved (quantities).

Now, one direction in which we expect to be able to push our theories is towards conceptions of 'underlying structure' which will allow the various manifest properties of each of the different kinds of some type to be explainable by reference to their characteristic underlying structures. Thus for instance we would expect lead's various observable features and its observable interactions with other substances all to be explainable in terms of its basic physical constitution. So in a sense our notion of 'characteristic underlying structure' is a notion of that sort of property which will account for all the manifest features of the kind in question. And in so far as our theories do seem to capture such underlying structures, then of course we will take the various different kinds of some type to be differentiated in essence from each other by their characteristic such structures.

Now what this means is that as we develop our ideas there will be an interaction between assumptions about what manifest properties show something to be an example of a given kind and assumptions about what underlying structure is characteristic of that kind. On the one hand, we will seek an idea of underlying structure which will account for the recognized clusterings of manifest properties; but, on the other, we will want to tailor our views on which manifest properties indicate membership of the kind to our existing conception of its underlying structure. This latter tendency leads to the rejection of such assumptions as that 'all that glisters is gold' (fool's gold), and 'all carbon is black' (diamonds). But the former kind of aspiration is of course what leads to our developing new ideas about underlying structures in the first place, as when we try to identify the molecular constitution of commonsensically identified compounds, or when we opt for nuclear charge rather than atomic weight as showing what counts as 'lead'. And so, more generally, it is the former kind of aspiration that leads to adopting certain 'kind identity principles' rather than others: we choose just those ways of differentiating kinds in terms of underlying structures which premise to give rise systematically to explanations of the currently admitted clusterings of manifest properties.

So there are two ways in which we can change our criteria for identifying examples of given natural kinds: we can alter our ideas of which manifest properties indicate membership of the kind, or we can alter our ideas of its underlying structure. Which kind of revision is to be made at any given juncture will depend on the relative methodological worth of the alternatives: is empirical progress promised better by adjustments in manifest criteria, or by a different account of underlying structure? As a rule the former alternative will offer itself as the more appropriate, for the theoretical revisions involved in changing manifest criteria will be far less central than those required by altered conceptions of underlying structure. (Remember, part of what makes them count as 'underlying structures' is that many generalizations about manifest features can be explained by references to them.) However, there will also arise occasions when more fundamental revisions are called for, when new ideas about underlying structure will in the end lead to a better account of the acknowledged concomitances amongst manifest properties. So it would be a mistake to think that ideas of underlying structure should always predominate over ideas about which manifest features indicate natural kind membership. Indeed if it were never appropriate to stick with existing ideas about manifest indicators, then there would never be any pressure to change our ideas about the underlying structures that differentiate natural kinds.

Yet it is precisely this thought, that assumptions about underlying structure should always predominate over manifest indicators, that is currently under examination. If assumptions about underlying structure did play a special role in fixing reference, then the referents of natural kind terms would always be guaranteed to be such as to make those assumptions true, and it would always be a mistake to reject them. This thought does, it is true, get some apparent support from the fact that it is usually appropriate to discount previously accepted ideas of manifest criteria when they conflict with assumptions about underlying structure. But, as has just been pointed out, it makes no sense to suppose that such discounting is always appropriate: if it were then nothing would ever lead to our developing new ideas of underlying structure.

The above arguments also show now that it would be a mistake to accord any special role to 'original samples' in fixing the reference of natural kind terms. The point of introducing 'original samples' was in a sense to get reference fixed independently of any accepted assumptions, apart from abstract assumptions about identity principles for the type of kind being named. In particular the 'original sample' story promised to preclude assumptions about the manifest features of a kind from playing any part in fixing the reference of the term for that kind. But we now see that in doing this the introduction of the notion 'original samples' simultaneously destroys any possibility of accounting for changes in our ideas about underlying structures and kind identity principles. If what a kind term referred to was just whatever shared 'underlying structure' with a given 'original sample', and if assumptions about how to recognize examples of the kind were never of any import in fixing this reference, then it would follow that whenever our view of underlying structures conflicted with such assumptions about recognizable criteria it would be the former view that was true and the latter assumptions that were false. If, for instance, ideas about greyness, heaviness, combinatory properties, etc. were really of no importance for what counted as a bit of 'lead', beside the requirement that it share 'underlying structure', according to some conception thereof, with a given particular lump of stuff, then we ought always to discount as examples of 'lead' any substances, however apparently similar, which are not, according to our current conception of 'underlying structure', of the same structure as that 'original' lump. But, as we have seen, we do sometimes change our ideas about 'underlying structures', precisely because doing so enables us to better account for the manifest properties of 'lead' (and the other material stuffs).

In any case, the idea of 'original samples' for natural kind terms has always been somewhat implausible. In particular, if the original sample is all that counts, and associated features are irrelevant for reference fixing, it is difficult to see how such a term could ever change its reference. But surely this ought in principle to be at least possible? In response to this problem some defenders of the causal approach have loosened the idea of an 'original sample', allowing in effect that the relevant sample can be expanded, by later actual applications of the term, and perhaps contracted, such as when it is decided that certain earlier applications did not fit the term after all. But then there arises the question of what governs these expansions and contractions. Presumably they result from accepted assumptions about how to recognize instances of the kind. At which point there seems no reason for not dropping the causal story altogether, and reverting to the holist view that, in so far as a given natural kind term refers to anything at a given time, it refers to just those things which have all the properties, both manifest and underlying, that current theory ascribes to the kind.

I shall conclude this section by reverting briefly to the causal theory as applied to proper names for people (or, more generally, to proper names for spatio-temporal particulars). It is not obvious that the criticisms relevant to the natural kind case apply to this case too. The arguments about the natural kind case stemmed from the possibility of changes in our kind identity concepts. We can and do change our minds about what makes two things the same material stuff, for example. It is not so clear that our concept of a person is similarly open to revision. Is it not uncontroversial that two 'stages' are parts of the same person just in case they are spatio-temporally continuous?

However there seems no reason in principle why our concept of a person should be absolutely inviolable. Consider the standard philosophical debates about personal identity. What would we say if we discovered that there were cases where certain spatio-temporally continuous bodies stopped having memories of the earlier circumstances of those bodies, and instead had memories appropriate to different bodies? Or what if we discovered cases where two given spatio-temporal bodies 'switched' external appearance and characteristic modes of behaviour? Why should we rule out the possibility of responding to such discoveries by deciding to apply our names for people to the 'entities' given by continuity of memories, or by continuities of appearance and behaviour, instead of to those given by simple spatio-temporal continuity? That is, why should we not allow that it might be appropriate to change our concept of a person, the better to accommodate the various characteristic memories, physical features, modes of behaviour, etc. that we currently associate with proper names for particular 'people'?

But if it is indeed right to allow this, then all the old arguments come into play again, with the implication that accepted assumptions about the manifest features of the bearers of particular proper names are as important in fixing the references of those names as any 'original samples' plus conceptions of person identity. For it is only if we recognize the importance of assumptions about manifest features that we will be able to give any account of the possible rationale for ever changing our conception of person identity.

There does however remain a relevant difference about the case of names for people and other spatio-temporal particulars. This relates to the possibility of 'qualitatively' identical people (towns, paintings, etc.) It is in principle possible that two or more people might look, behave, remember, and so on, in exactly the same ways. Yet they would still be different people, and it ought to be possible for a specific proper name to refer to the one rather than the other. So it cannot be that the reference of a proper name depends only on the descriptive features associated with that name. To allow for the name's uniqueness of reference in such cases we want to allow that the user's relation to particular samples, or 'stages', of the particular person in question plays some part in fixing the name's reference. An account of this 'indexical' aspect of proper names might be given in terms of the 'original samples' story, or in terms of some other type of quasi-demonstrative convention. However, the point made in the last paragraph still stands: we still need to think of the specific descriptive facts associated with a name as also significant for fixing reference, lest we preclude any account of what might lead to alterations in our concepts of persons and suchlike. The difference between this case, and the case of natural kind terms, where 'original samples' were argued to play no role whatsoever, can be understood as due to the general nature of conceptions of the identity of spatio-temporal particulars, as opposed to conceptions of natural kind identity. Abstracting from such conceptual alterations as we have just envisaged, our concepts of spatio-temporal particulars allow that two particulars can be identical in all respects (bar spatio-temporal location) yet distinct. So the use of a name to refer to a specific such particular requires that use to be somehow spatio-temporally related to some exemplification of that particular. But there is no possibility of distinct natural kinds sharing all 'qualitative' features: a substance that was in all respects like gold, say, would just be gold itself. Which is why there is no corresponding need for 'original samples' or anything similar to help fix the reference of natural kind terms.

What now about the earlier observation that the causal theory's discounting of the descriptive features associated with a proper name seems to fit well with our intuitions about the use of such names: (Recall the 'Bertrand Russell'—*Philosophical Investigations* example.) Here we need to bring in Putnam's useful hypothesis of the *division of linguistic labour*. (See Putnam [1973a].) Putnam suggests that in our use of many terms we defer to 'experts': we allow that what such terms refer to depends not on the ideas, such as they are, which we as individuals associate with those terms, but on how the *cognoscenti* use those terms. Take the term 'gold', for instance. As Putnam points out, very few people really know what in the end shows something to satisfy this term. Yet everybody's judgements involving this term are answerable to what a 'special sub-class of speakers' (p. 703) would count as satisfying it. A similar claim could obviously be made about 'Bertrand Russell'—what this name refers to depends not on the individual speaker's ideas but upon those of people who are acknowledged to know about such matters. But this illuminating thesis of the division of linguistic labour should not be considered, as it often is, as any argument for the causal theory of reference. For it is perfectly consistent with any account of what makes a term as used by experts refer to what it does. In particular, it does not show that the descriptive facts that expert speakers associate with a given term are irrelevant to what that term refers to, even if unexpert speakers' such associations are so irrelevant.

6

LOGICAL FORM, HOLISM, AND TRANSLATION

1 AN ARGUMENT AGAINST HOLISM

Dummett has in a number of places objected to holist theories of meaning on the grounds that they preclude any systematic account of the mechanisms which govern the acceptance and rejection of statements. (See in particular his [1973], pp. 623–7, and his [1974a], pp. 392–7.) The meaning of a sentence depends on the meanings of its constituent expressions. If the meanings of the expressions in a language depended in turn on the totality of sentences accepted by the users of the language, then we would never be able to understand any decision to accept any sentence as being guided by a prior grasp of what that sentence meant. But surely at least some such decisions are to be so understood. From which it follows that at least some of the sentences in a language must be composed of expressions the meanings of which are fixed prior to and independently of decisions as to whether those sentences are to be accepted.

How does this argument apply to the 'holist' account of the meaning of scientific terms I have defended? It is relevant here that my holism is less than total. I have not argued that the meanings of scientific terms depend on *all* the accepted sentences containing them, but only on the surrounding context of accepted *lawlike generalizations*. So there need be no special difficulty about explaining how decisions on other kinds of sentences than generalizations can be informed by a prior grasp of their meanings. In particular, as I shall have cause to observe further below, there is no such difficulty for singular sentences about spatio-temporal particulars.

But this point scarcely removes the sting from Dummett's objection. For I am still left with a puzzle about the lawlike generalizations themselves. Surely in at least some cases our decisions as to whether to accept lawlike generalizations must be informed by a prior understanding of what they mean? But, again, how is this to be possible, if

the meanings of their constituent terms depend in turn on whether
they are accepted or not?

Dummett does, it is true, allow that we need to move some way
towards holism. He favours a 'molecular' view of language rather than
a strictly 'atomistic' account. That is, he allows that there are some
sentences which play a role in fixing meaning, that for some sentences
the meanings of the constituent expressions depend on those sentences being accepted rather than vice versa. (Cf. Dummett [1974b],
p. 19.) But then of course the awkward question of what enables us to
decide the acceptability of these sentences does not really arise: it is a
consequence of their special role in the language that their acceptability cannot be at issue. And this means that Dummett's molecular
concession to holism does not help with our problem. For the acceptability of lawlike generalizations in science can certainly come into
question. Indeed it is a central plank of my holism that *any* accepted
lawlike generalization in science is in principle capable of revision.

The difficulty we face here is essentially the same one as was raised
when the idea of a holist account of meaning was first introduced at
the end of Chapter 4. I pointed out there that on any traditional
conception of meaning the meaning of any non-analytic sentence
somehow specified what evidence would warrant accepting or rejecting that sentence—the actual acceptance or rejection of the sentence
then being conditional on the occurrence or non-occurrence of that
evidence. But, it was observed, on a holist conception of meaning this
could not work for generalizations: if the meaning of the predicates in
a generalization depended on whether that generalization was
accepted or rejected, such acceptance or rejection could not simply be
a matter of responding to the evidence specified as relevant by the
sentence's stable meaning. At that point I suggested that there need
be no real incompatibility between a holist conception of meaning
and evidential rationality, provided we gave up the assumption that
all meaning changes were *ipso facto* irrational. But my remarks at that
point did not explicitly deal with the more specific question of how
our understanding of the way the content of a generalization derives
from the meanings of its component expressions manages to inform
our evidential assessment of that generalization. I turn now to this
question.

2   MEANING AND LOGICAL FORM

On any account the meaning of a generalization will presumably depend not only on the predicates it contains but also on the logical constants therein; that is, on its logical form. In a manner of speaking the logical constants contribute the general shape of meaning for generalizations of that form, and the specific predicates involved will then fill in its distinctive content.

To give an account of the way the logical constants play this role we need a 'semantics of logical form' of the kind discussed in the last section of Chapter 3 above. Suppose we adopt the standard 'classical' version of such a semantics. (The issues at hand would arise in pretty much the same way for the semantics which derives from a verificationist approach to meaning. But as I shall be reconfirming the desirability of a 'truth conditions' conception of meaning below, a discussion in terms of the classical semantics will suffice.) The classical account of how the logical constants in a generalization contribute to its meaning combines the standard model theoretic explanation of variables and quantification with the truth-tabular account of the truth functions. To take one of the simplest cases, this tells us that a generalization of the form (x) (Fx $\supset$ Gx) will be true just in case everything that satisfies F also satisfies G; that is, if F's extension is contained in G's. This goes for any generalization of that form, whatever the actual F and G. The specific meaning of any actual such generalization will then be 'filled in' by a specification of what determines the actual extensions of the predicates which stand in place of our schematic F and G. And a similar story can be told, recursively, for more complicated generalizations, up to any degree of complexity.

The problem at hand now takes the following form. To understand the meaning of a generalization is, we are supposing, to understand what would make it true. But to understand that we need to have a grasp of what determines the extensions of its constituent predicates. The holist theory of meaning, however, implies that the extensions of predicates are not fixed independently of the totality of decisions as to which generalizations containing them are to be accepted. (Indeed because of this it implies moreover that scientific predicates do not have determinate extensions at all.) But how then can it be, as in at least some cases it must, that decisions on generalizations derive from a grasp of their meanings?

This apparent paradox will be resolved in the next section. But before that it will be useful to comment at length on some subsidiary issues. First I shall discuss whether the apparent unrevisability of 'logical truths' calls for some modification of our holist account of meaning. Then I shall consider how far such truths are indeed unrevisable, and this will then lead on to some general remarks about semantic accounts of logical form.

At the end of Chapter 3 it was pointed out that a semantic account of the workings of the logical constants will have the effect of guaranteeing all sentences of certain forms (and, correspondingly, certain inference schemata). Thus, for instance, the classical semantics of quantification will guarantee that any sentence of the form (x) $\sim$ (Fx. $\sim$Fx) is true: since the extension of $\sim$F is the complement of the extension of F, no individual can satisfy both F and $\sim$F. (A verificationist-type semantics would correspondingly guarantee the *assertibility* of sentences of certain forms.)

Now the possibility of such guaranteed sentences might seem to be in tension with the arguments for a holist account of the meanings of scientific terms. The primary motivation for this account was that there seemed no privileged sub-set of generalizations which were constitutive of their constituent expressions' meanings and hence in principle unrevisable. In essence the argument was that since no generalization could be singled out as distinctively meaning-constitutive we were constrained to suppose that all accepted generalizations played a part in constituting meanings. But now it seems that there are some generalizations of the required kind after all.

However it is important here that the holist theory of meaning I have defended is intended specifically as an account of the meaning of scientific *predicates*. And as such it is in no real tension with the privileged status of logically guaranteed generalizations. For such generalizations cannot be considered to play any significant part in fixing the meanings of the predicates involved. Such generalizations are guaranteed to be true precisely because they will be true whatever specific predicates enter into them. Such special commitment as is owed to them derives from the meanings of their logical constants alone and not from the meanings of the predicates involved. (Except in so far as it can be said that the *form* of predicate meaning, the meaning common to all predicates by virtue of their being predicates, is given by a semantics of logical form.) So when it comes to the question of what gives a specific predicate its specific meaning, what differentiates it in meaning from other predicates, we still have to attend to the total structure of accepted generalizations containing it. And this means that the earlier arguments against taking any privileged sub-part of this structure to play a special role in fixing predicate meanings still stand in favour of a holist account of those meanings.

I turn now to the question of how far the logical truths are indeed unrevisable. Why exactly should such statements be considered inviolable? Notwithstanding any special status they may have, logically true generalizations are still part of our theories of external reality. So why should it not sometimes be that their revision is the best way of dealing with some set of empirical anomalies? Thus for instance, it has been argued that the best way of accommodating certain difficulties raised by quantum mechanics is to abandon certain standard logical principles (cf. Putnam [1969]). But if such revisions are allowed, then in what sense do the semantic workings of the logical constants 'guarantee' the logical truths?

So far I have been going along with Dummett in supposing that we need to account in semantic terms for the acceptance of sentences in which only the logical constants of the propositional and predicate calculi occur essentially (that is, for those sentences such that any other sentences of the same form with respect to those constants are also accepted). But why should we suppose this? Why should we not simply be able to decide to accept all sentences of a certain form on grounds of convenience or on general methodological principles? Why for instance should we not simply be able to resolve that henceforth we are going to accept all sentences of the form p v ~p, without being able to give any semantic theory which validates this decision? Or, conversely that we are not any more going to accept all sentences of the form [p. (q v r)] ⊃ [(p. q) v (p. r)] (which is what the suggestion made in connection with quantum mechanics amounts to)?

Indeed Dummett's view that the logical truths are forced on us by the semantics of the logical constants is by no means unquestioned orthodoxy. There is a significant school of thought, with Quine the foremost representative, which takes the 'logical truths' to be substantiated in just the same way as any other highly stable propositions. Exactly what way that is—whether it is a matter of conventional stipulation or of centrality in our empirical theories, and whether correspondingly their unrevisability should be thought of as absolute or relative—depends on what view is taken of the analytic-synthetic distinction. But either way we have an alternative to Dummett's view that 'logical truth' has to emerge from an independent semantic account of how the logical constants function. (The scare quotes here are in recognition of the circumstance that on the anti-Dummettian view there is no clear dividing line between 'logical' and other general truths. However to avoid awkwardness I shall drop them henceforth.)

We can bring out the point at issue by considering once more the debated law of the excluded middle. The question to be put to Dummett is: Why should the acceptance of this principle be answerable to an independent semantic account of the way 'v' and '∼' function in producing the meanings of complex sentences from simpler ones? Why should it not simply be accepted straight off, even if it cannot so be accounted for (as would be the case if we opted for a verificationist approach to meaning)? (Cf. Wright [1976], pp. 240–4.) We would still of course want to give some account of the way 'v' and '∼' generally operated in making complex sentences from simpler ones, for we would still want to give a systematic account of how meanings are generated for the general run of such complex sentences. But the acceptance of the principle of the excluded middle could then be considered to play a further part in fixing the meanings of 'v' and '∼' over and above what is given by the verificationist analogue of the truth tables.

At this point it will be helpful to distinguish two ways in which a semantics of logical form could be supposed to 'account for' logical truths. In the first place such a semantics might be taken to *explain* why a given linguistic community has as a matter of fact adopted a certain logical practice. Alternatively, it might be taken to *justify* that practice.

To see this, let us consider in more detail the nature of such a semantics. Such a semantics specifies how meanings of certain complex sentences derive from their structure with respect to certain kinds of constituent expressions. Now, because they have the meanings they do, those complex sentences can be used to make certain claims about reality. And this means that in showing how those meanings are systematically made up our semantics will show how the use of those sentences presupposes certain kinds of relationships between the constituent expressions and elements of reality. Thus the standard semantics of the propositional calculus shows how the working of the truth functions requires each sentence to have a meaning such that reality will make that sentence either true or false but not both. And the standard semantics for the predicate calculus shows similarly how quantification presupposes that every singular term has a meaning which makes it stand determinately for some individual (or determinately for none), and that every predicate has a meaning which makes it stand determinately for a class of individuals. (So far I have been considering only sentence structure in relation to the logical constants of the propositional and predicate calculi. But other kinds of iterative operators also give rise to sentence structures relevant to the meanings of complex sentences. And so we can have semantic accounts of these structures too, which will similarly suggest relationships between the constituent expressions and aspects of reality. Thus there is the possible world semantics for modal discourse, and similar possible world accounts of sentences about propositional attitudes. The general remarks in the remainder of this section will be applicable to these semantic theories too. But for simplicity I shall continue to concentrate on the logical constants of the propositional and predicate calculi.)

So, to repeat, a semantic theory of the kind we are considering is a theory of the way the contents of certain kinds of complex sentences derive systematically from the contents of their constituents. In capturing that systematization such a theory shows how the use of those sentences presupposes certain relations between those constituents and reality. Now, we can ask two questions, given such a theory of a certain aspect of a given community's linguistic practice. Firstly, is the theory right about the way those people understand the complex sentences in question and so conceive of their constituents' relation to reality? Is it indeed true, for instance, that they understand 'and', 'not', etc., in the way indicated by the truth tables and so presuppose that every sentence to which these operators can be applied is either true or false but not both? Do they understand quantification in the manner implied by the standard model theoretic account and therefore think of every predicate as having a meaning which gives it a determinate extension? Then, secondly, assuming that the semantic theory is right about how the people in question understand the complex sentences under examination, we can ask whether they are right to have complex sentences which they understand in that way. Or, more illuminatingly, are they justified in presupposing that the constituents of those sentences relate to reality in the way required by that understanding? Are they indeed right to suppose that all their sentences (any of which can function as constituent expressions in truth functional compounds) have contents such that reality will always be such as to make them determinately true or false? Do their predicates indeed have meanings which ensure determinate classes of individuals are picked out as their extensions?

It should now be clear why there are two senses in which a semantics might 'account for' the logical truths and inference patterns which arise from a certain kind of sentential complexity. In the first place it might simply be intended to *explain* why as a matter of fact a certain community accepted those truths and inferences. In this case the crucial question would be whether the semantics accurately reflected the way that community derived the meanings of the complex sentences in question and thus the way they took the constituent expressions to relate to reality. As such, the semantics would then show, as a kind of by-product, in the way outlined before, how certain 'guaranteed' truths and inferences were forced on them. That is, certain special sentence forms would be shown to be understood in such a way that it would seem to the users of those sentences that reality could not but be as those sentences portrayed it. But then, secondly, a semantic account might be intended to *justify* certain logical truths and inferences. And in this connection the crucial question would not be how the community in question understood the sentential complexity under investigation, but whether they were entitled to such an understanding.

As suggested above, it might well be that the constituent elements in those complex sentences did not in general succeed in relating to reality in the way presupposed by their understanding of those sentences. What this would mean is best brought out by switching to the case where the linguistic community in question is our own. Suppose we concluded, on the basis of arguments in the philosophy of language, that the way we had up to now understood certain complex sentences required the constituent expressions to relate to aspects of reality in a way they in fact failed to do. We would then no longer feel that we could maintain our previous understanding of those sentences. Of course we could, and in most cases would, seek some alternative way of understanding them, based on an improved conception of how their constituents related to reality. But it may well then be that we ended up with a rather different set of 'guaranteed' truths and inferences for those sentences. For there would be no assurance that our new understanding of the compounding operations involved and the constituents they had to work on, would indicate the same sentence forms to be unimpeachable representers of reality. And in such a case we would *not* take the 'semantic theory' of our previous understanding of the sentential complexity to have *justified* our previous acceptance of certain statement forms as logical truths, even though it explained it all right.

(This distinction between two ways in which a semantics might 'account for' certain logical truths can usefully be applied to the assessment of the possible world semantics for modal discourse. This may well be an accurate reflection of the way we understand such discourse, and as such it would succeed in explaining why we happen to accept certain 'guaranteed' truths and principles of inference involving modal terms. But it does not follow that we would be entitled to have such a possible world understanding of modal discourse. Indeed surely we would not be, for it would commit us untenably to a full-bodied realism about possible worlds. This is not to deny the possibility of our developing, or already having, a different semantic understanding of modal discourse, which *would* entitle us to roughly the same set of truths and inferences as we now accept. 'Possible world' accounts of statements about propositional attitudes are a different matter, for those 'possible worlds' can be understood realistically as in some sense being in the minds of the individuals to whom the attitudes are ascribed.)

Let us now return to the question of whether the acceptance of logical truths has to be accounted for by reference to a semantic theory. We can now see there are really two questions here. First, does a community's accepting certain logical truths always demand *explanation* in terms of their understanding of the relevant kind of sentential complexity? And, secondly, are they *entitled* to those logical truths only if they emerge from a legitimate such understanding?

Taking the second question first, there is of course one sense in which we are quite free to accept all sentences of a certain form with respect to particles like 'and', 'not', 'all', etc., if we want to, without giving a justification in terms of an independent account of the way those particles generally work in producing complex sentences. After all, they are only words, and there is nothing to stop us saying what we like. However, as Humpty Dumpty makes clear ('there's glory for you'), there is another sense in which we are not free to say what we like. Language works only because to a large extent words have unitary meanings governing their use across different sentential contexts. If words always shifted their meaning from context to context there would be no possibility of our understanding a potential infinity of sentences in terms of the way they were generated from a finite number of constituents and compounding operations. Now, the part that an iterative operator plays in generating a potential infinity of sentences must be understandable in terms of the way those resulting sentences' claims about reality depend on certain supposed relations between the constituent expressions and elements of reality. And this means that a decision to accept all sentences of a certain form in disregard of such a semantically specifiable meaning for the operators involved will always be open to an accusation of trivial equivocation. However, what exactly is it for a word to have a 'unitary' meaning? There is no doubt that a semantically specifiable meaning for an iterative sentential operator is so unitary. But does it have to be mere equivocation to make additions to, or place restrictions on, the meaning which would be yielded for an iterative operator by such a specification alone? Why can we not think of *the* meanings of 'v' and '~' as being given by the verificationist account of how they work as sentence formers, plus the stipulation that all the sentences of the form p v ~p are unconditionally assertible? Of course such a situation would require us to recognize that the meanings of the particles in question were not after all exhaustively specified by the way they generally worked to produce complex sentences. But provided we did recognize this, and were careful to use language accordingly, it is difficult to see what would be wrong in any absolute sense. Nevertheless it is undeniable that such an understanding of an iterative operator adds complexity to the language. And this in itself is a strong reason for resisting the imposition of such understandings. I do not want to say that there could never be some arguments in favour of such an imposition. But I am inclined to be sceptical about whether they will often suffice to overcome its intrinsic undesirability.

However, we have now seen that there is another rather different sense from that first envisaged in which logical truths might be revised: namely when we alter logical truths, not in defiance of an existing semantic explanation of the operators involved, but in accordance with a revised semantic understanding of those operators. The case considered earlier was where we decided, on the basis of considerations in the philosophy of language, that we were not in fact entitled to our existing semantic understanding of certain sentence structures, that the constituent expressions could not relate to reality in the way required by the understanding. It was pointed out that this could lead us to seek a new understanding of those sentences, which could then result in a rather different set of guaranteed truths. But it is also possible that we could look for a new understanding precisely in order to end up with a different set of guaranteed truths, which promised to cohere better with our scientific theories. In this case the motive for the revision would come, not from reflection on the possible relation between language and reality, but directly from reflection on non-linguistic reality itself.

Now, what about the *explanation* of a community's acceptance of given logical principles? Does this have to be in terms of their grasp of the relevant sentential complexity? Well, I allowed above that there is nothing to make it absolutely illegitimate to adopt logical principles without a semantic substantiation. So, *a fortiori*, it cannot be absolutely impossible; and there can be no absolute requirement that the explanation of a people's adopting a logical principle should always proceed by showing that they have a semantic theory which drives them to it. It may just be that they have simply accepted it as such without a semantic substantiation. On the other hand, we would certainly expect that in the great majority of cases such a semantic explanation will be the right one. And in any case, whenever we can explain a community's adoption of a certain body of logical truths by attributing a certain semantic theory to them, then surely we should. Why should we look a gift explanation in the mouth?

Now this is all relevant to the issue discussed at the end of Chapter 3, of whether a people's acceptance of logical principles could show us what model of meaning was appropriate for the constituent expressions of the relevant sentence structures. To get properly clear on this issue we again need to separate out two questions. First, there is the question of whether their accepting certain inferential principles for sentences of certain structures shows us how they understand those sentences, and therefore how they conceive of the meanings of the constituents. Then there is the further question of whether they are entitled to their understanding of those sentence structures, and their conception of the constituents' meanings, and whether they are consequently justified in ending up with the logical principles they have.

Dummett maintained that since people can err logically we cannot take their inferential practice to show us anything about how the sentence structures involved are to be understood. We are now in a position to appreciate this argument properly. If anything, what their actual inferential practice can show us is how *they* understand those sentence structures. But what it cannot show is that they are *entitled* to so understand them and to their model of meaning for the constituent expressions.

There are perhaps reasons for doubting whether a people's inferential practice can even show us how *they* understand the relevant sentential complexity. Might they not fail to see that their understanding of the sentence structures in question commits them to the validity of certain inference schemata, or think that it does when it does not? It is not implausible that such mistakes could occur in connection with sophisticated sentence structures or where the inferential principles in question are complex ones. But it is scarcely likely in the cases relevant to the discussion of Chapter 3, namely, sentence structures with respect to the truth functional connectives and such inferential principles as go directly with acceptance of the law of the excluded middle (like double negation elimination, $\sim\sim p \vdash p$). Then there remains the abstract possibility that the inferential principles under consideration are accepted straight off by the people in question, and do not in fact flow from their semantic understanding of the related sentence structures. But, as pointed out a couple of paragraphs earlier, that we can explain a people's inferential practice by attributing a certain semantic understanding to them is surely every justification for doing so.

However, our ability to use people's inferential practice to deduce how they understand the relevant sentential complexity leaves untouched the possibility that they might be in error in having that understanding in the first place. That is, it leaves it open that they are mistaken in taking the elements of that structure to relate to reality in the way their understanding requires. In particular, the fact that people accept the forms of reasoning systematized in the classical propositional and predicate calculi does not show that Dummett is wrong to doubt that the sentences to which the truth functions get applied (that is, all sentences) in general have meanings such that reality always makes them either true or false, nor does it show I am wrong to doubt that predicates in general have meanings such as to give them determinate extensions. All it shows is that people think these things.

Dummett inclines to the view that the only coherent model of sentential meaning is one which has sentences getting meanings from their association with verification conditions. And this leads him to suggest that we ought to, even though we do not, withhold unqualified assent from the law of the excluded middle: on the verificationist picture the meanings of p and ~p will not in general be such as to ensure that reality will make either the one or the other verified. But in Chapter 3 I pointed out that a similar line of reasoning suggests that we ought also to withhold assent from the principle of non-contradiction. For the verification conditions we attach to p and ~p, for many p, seem to make it possible, and indeed often actual, that both p and ~p are verified. Are we then to say that our acceptance of the principle of non-contradiction is an error stemming from our adoption of the incoherent model which pictures our sentences as getting meaning from associations with truth conditions rather than verification conditions? Surely not.

(A defender of the verificationist view might at this point be inclined to object to my contention that the verificationist model of meaning fails to validate the principle of non-contradiction. Does not the verificationist explain '.' as giving rise to sentences p . q which are verified just in case both p is verified and q is verified, and does he not explain '~' as giving rise to sentences ~p which are verified just in case it is verified that p is not verifiable? And does it not follow from these explanation that a sentence of the form p . ~p can never be verified? Indeed it does. But what now comes in question is the verificationist explanation of '.' and '~', and in particular of '~'. Since we know that procedures for verifying p and ~p are often such that reality turns out to allow both p and ~p to be verified, surely we are not *entitled* to understand '~' as in general giving rise to sentences ~p whose verification amounts to a verification of the unverifiability of p.)

There now seems to be a strong case for a truth conditions model of meaning. It is surely right to think that the principle of non-contradiction is susceptible of a semantic justification. The verificationist account of meaning cannot provide this. But if we suppose that the contents of sentences are given by concepts of truth conditions which transcend accepted verification conditions, then there is no barrier to such a justification. That accepted methods for verifying p and ~p can on occasion lead to both being 'verified' does not show that the sentences in question have contents such that reality can make both true, but merely that their contents are not adequately represented by their verification conditions, and in particular that our adoption of verification conditions is not faithful to our understanding that what makes ~p true is the absence of what makes p true.

But how, from my point of view, can we be justified in supposing that our assertions have verification-transcendent truth conditions of the kind required for us to be entitled to understand ~p as having a meaning which makes it true whenever p is not? For at the level of the sentence structure generated by operations on predicates this requires that we understand any predicate ~F as having a meaning which gives it an extension which is the complement of F's extension. Yet I have argued throughout that the contents we attach to predicates are not such as to pick out determinate classes of individuals as their extensions. If a given predicate F does not have a determinate extension, then how can we be entitled to understand ~F as the predicate whose extension is the complement of F's (non-existent) extension?

The solution to this puzzle is to recognize that our grasp of predicates, and consequently of assertions, contains a *programmatic* element. Take a pair of predicates like 'brave' and 'not brave', or 'soluble' and 'not soluble'. The distinctive contents we attach to 'brave' and 'not brave' are in effect identified by the various things we recognize to demonstrate 'bravery' and 'non-bravery'. These criteria are indeed not such as to ensure either that everyone will be shown to be either 'brave' or 'not brave', or such as to ensure that nobody will be shown to be both 'brave' and 'not brave'. But in addition it is part of the notions we attach to this pair of terms that eventually our criteria for recognizing 'bravery' and 'non-bravery' will be developed to such a point that these things are ensured. This programmatic faith is premissed on a supposition that there is some kind of underlying state, such as a certain physiological structure, such that our 'eventual' use of 'brave' will make it apply to just those individuals who have that state, and our 'eventual' use of 'not brave' will make it apply to just those individuals who lack that state.

To recognize that the contents we attach to our predicates contain such a programmatic element is not to deny the earlier arguments for doubting that predicates have determinate extensions. Our commitment to there being some underlying state to which 'brave' is in the indefinite long run going to attach, in combination with the criteria we currently associate with 'brave', need not in any serious sense pick out any determinate class of 'brave' individuals. There may in actuality be any number of different states which fit the bill to a greater or lesser extent, and there need be nothing in our present use of 'bravery' to make it stand for one rather than others. Even if we suppose that in the infinitely distant 'perfect' theory the term 'brave' ends up standing for a specific one of these states, it does not follow that that state is unequivocally picked out by our current use of that term. In all likelihood it will be equally faithful to our current use of 'brave', to the criteria that currently give this term its distinctive content, to say that it stood variously for different of the candidate states, or that it did not really stand for any real state at all.

But if its current content so fails to give 'brave' a determinate extension, in what sense are we entitled to our 'programmatic' supposition that it has one? After all, what is at issue is whether we can have an understanding of the truth functions which will *justify* the principle of non-contradiction, and what matters for this is not that we in some sense *do* think of our predicates as having determinate extensions, but that we *should* so think of them. However the important issue here seems to me not so much whether our current use of specific predicates is such as to attach them unequivocally to underlying states which will give them determinate extensions, but whether we can in general legitimately aim at such attachments. And given that there are such states, then surely this aim is legitimate. That this is the right way of conceiving the matter is best shown by considering what happens when we abandon such an aim for a given predicate. Suppose, for instance that at some time our current criteria for 'brave' and 'not brave' came to appear so fragmentary and incoherent as to make us lose our programmatic faith and decide that after all there was no underlying state for 'brave' to attach to and for 'not brave' to attach to the absence of. Such a decision would not, and should not, result in our concluding that our predicates in general did not have the kind of meaning that is required for us to understand the truth functions in the way that justifies the principles of non-contradiction and the excluded middle. What we would do is conclude that '*brave*' did not have the kind of meaning that is required for us to understand 'not brave' as standing for a complementary extension. That is (supposing that we continued to use the terms 'brave' and 'not brave') we would not take the 'not' involved seriously, and would cease to understand these terms as either contradictories or contraries: we would simply give up the expectation that everybody satisfies 'brave' or 'not brave' but not both.

So the legitimacy of the programmatic element in our grasp of predicates and the assertions they enter into, by which we are entitled to a classical understanding of the truth functions, does not depend on any specific predicate and its negation fulfilling their programme. If a specific such pair are deemed incapable of keeping their programmatic promise we simply withdraw the classical understanding of negation for that case, while continuing to keep it for the general run of pairs of predicates which we are still working to give appropriate extensions to.

(It might seem that a verificationist model could be similarly defended by appeal to programmatic elements of meaning. My argument was that a legitimate verificationist explanation of negation could not guarantee that sentences of the form p . ~p would never be assertible. But why should a verificationist not argue that even if we do not now do so, it is nevertheless part of our grasp of sentences that we will develop verification conditions for them which will guarantee non-contradiction, but will not exclude middles? The problem with this defence is that it is very difficult to see what possible rationale there could be for such hopes other than the expectation that (some) pairs of predicates and their negations will eventually get attached respectively to the presence and absence of underlying states. If we did not expect that there was some underlying state for the verification conditions of 'brave' and 'not brave' to get right eventually, then why should we think that their verification conditions will ever be guaranteed not to apply simultaneously? Why should 'not bravery' ever establish the impossibility of 'bravery' unless it were that it had come to mean the absence of what made people 'brave'? So the suggested programmatic version of verificationism is in the end no different from a model of meaning in terms of (programmatic) truth conditions.)

3 HOLISM AND LOGICAL FORM

We can now finally return to the paradox facing our holist theory of meaning: how can decisions on generalizations be informed by a grasp of their meanings, if those meanings depend in part on whether those generalizations are accepted? Or, as it came out more specifically, given the standard semantic account of the logical form of generalizations, how can we understand decisions on generalizations as being informed by an understanding of what determines the constituent predicates' extensions, if what that is depends on what such generalizations are accepted?

Now, the holist theory of meaning is indeed right to imply that decisions on scientific generalizations cannot be characterized in the simple terms initially suggested by the standard semantics: it is indeed not possible for scientists to first grasp what determines the extensions of their predicates, and then, on that basis, come to conclusions about which generalizations the evidence supports. But the remarks at the end of the last section show that we need not conclude that the matter cannot be seen in the terms suggested by the standard semantics at all—for the circle involved can fortunately be encompassed as a whole.

Suppose we see our semantics of the logical form of generalizations as telling us simply that we ought to accept generalizations the extensions of whose predicates are such as to make those generalizations true. Why should it be any obstacle to our pursuing this aim that the accepting of a system of generalizations and the fixing of their constituent predicates' extensions should proceed side by side? From this point of view, our aim is developing a scientific theory is a system of generalizations which, together with the way the world actually is, will give our predicates extensions such that the truth conditions of those generalizations, as given by the standard semantics of their logical form, will be satisfied. So our understanding of the way the meanings of generalizations depend on the meanings of their constituents, our understanding of their logical form, does inform our decisions as to whether to accept or reject them. But what is important is that this is not a two-stage process where predicates are somehow first assigned extensions, and then our understanding of how that assignment is made tells us how to decide on generalizations containing those predicates. These two processes have turned out to be different sides of the same coin.

It is of course unlikely that we will ever actually have a system of generalizations which will determinately yield extensions which are such as to make all those generalizations true. Our theories will always face anomalies, and this will make their predicates' extensions indeterminate and the application of the notion of truth (or falsity) problematic. But we can still see science as aiming at the goal of truth. Just as in the last section I justified the assumption of determinate extensions programmatically, so here we can see the goal of truth, which requires such extensions, in the same way.

On the normal view, the elaboration of scientific theories does involve a kind of two-stage process. First of all certain generalizations, and the observational and inferential procedures they underpin, are picked out as playing a special role in fixing the contents of the terms they contain; and this then indicates what evidence is appropriate for or against other generalizations containing those terms. The former generalizations, because of their special role, are self-validating; the latter are accepted as and when the indicated evidence is available. However, there is no such distinction amongst scientific generalizations. As we have repeatedly discovered, we have no alternative but to suppose that all scientific generalizations contribute to the fixing of the contents of scientific terms. So there is never any question of evidence for or against a single generalization in isolation. A 'counter-example' in the form of an anomaly always calls in question not just the generalization to which it is a 'counter example' but also the generalizations behind our accepting it as such in the first place. Correspondingly, inductive evidence accrues not to single generalizations but to our theories as wholes: the accepting of a positive instance for a given generalization will always presuppose other generalizations, in such a way that it is the whole set of generalizations involved, if anything, that past evidence supports. This holist approach to scientific theories has already, in earlier chapters, been shown to be consistent with the notions that our theories are attempts to represent external reality, and that we can have good reasons for thinking one such representation better than another. What has in addition been shown in this chapter is how our understanding of the internal semantic *structure* of generalizations plays a part in our development of our theories.

The contribution made by this understanding of semantic structure in fact comes out most clearly in connection with our appreciation of the logical relations between different statements. We hope in eventual prospect to develop a system of generalizations which are all determinately true. Since the same predicates come up in different statements, and since we understand the (prospective) truth conditions of those statements as depending on the (prospective) extensions of their predicates, this imposes certain restrictions on the combinations of statements it makes sense to accept. Thus if we accept $(x)\,(Fx \supset Gx)$ and $(x)\,(Gx \supset Hx)$, we cannot accept $(x)\,(Fx \supset {\sim}Hx)$. It is our grasp of semantic structure which explains why this makes no sense. Not that it would be correct to say that our predicates *have* extensions such that these three generalizations cannot all be true. Rather it is that if we were prepared to accept such a combination of generalizations we would have no chance of *giving* our predicates extensions which made those generalizations all true.

4 QUINE AND INDETERMINACY

The rejection of the traditional conception of scientific development as a kind of 'two-stage process' stems of course from the work of Quine. Quine denies that there is any substance to the distinction between privileged 'analytic' generalizations which lay down the meanings of the terms they involve and 'synthetic' ones which are then answerable to the evidence indicated by such meanings. However, this denial leads him not to a holist theory of meaning but to his thesis of the *indeterminacy of translation*. Quine argues that:

> . . . manuals for translating one language into another can be set up in divergent ways, all compatible with the totality of speech dispositions, yet incompatible with one another. In countless places they will diverge in giving, as their respective translations of a sentence of the one language, sentences of the other which stand in no sort of equivalence however loose. ([1960], p. 27.)

This is not the conclusion we get from a holist theory of meaning. Assuming that translations should preserve meanings, the holist theory of meaning implies, not that there will be many different translations of the language of one scientific theory into that of another, all equally good, but that there will be *none*. If the meanings of the terms in the respective languages depend on the total structures of the theories involved, then those terms will not share meanings and no translation will be possible. Unless, of course, those total structures are the same and the 'two' theories are really one, as is no doubt the case with, say, the theory of contemporary German-speaking physiologists and that of their English-speaking counterparts. But then, again, there will not be many translations, but just *one*, namely that which pairs each German term with the English term that shares the same place in the theoretical structure in question.

In *Word and Object* Quine gives a special status to the 'stimulus meanings' of sentences. (The stimulus meaning of a sentence is the ordered pair of those sensory stimulations that would prompt assent and those that would prompt dissent.) Quine takes stimulus meaning to be 'the objective reality that the linguist has to probe when he undertakes radical translation' ([1960], p. 39). So what he requires of a translation is that it preserves everything that stimulus meanings can tell us about the use of sentences. But then he argues that we can preserve stimulus meanings equally well by a number of different 'analytic hypotheses', by a number of different translation manuals, each of which will give a different English sentence for a given foreign-language sentence.

From the point of view developed here (and indeed from Quine's point of view in his [1951]), special treatment of stimulus meanings seems unwarranted. For, as we have seen, the stimulus meanings of words are no more authoritative and stable than other features of their use. But we can let this pass for the moment. For what is far more puzzling is why Quine takes it that stimulus meanings exhaust 'the totality of speech dispositions'. As a number of writers have asked, what has happened to the *inferential* connections between sentences? We nowadays recognize (not least because of Quine's own writings) that it is an essential feature of language that sentences are often asserted on the basis of other sentences, rather than in direct response to observation. But surely then the characteristic tendencies of language users to make such inferences are part of their 'speech dispositions'. And as such should they not also be preserved in translation? (Cf. in particular Boorse [1975], p. 371.)

We need to take some care here. Quine does require that translation preserve stimulus-analytic sentences, those sentences such that no stimulus would prompt dissent from them. And these he takes to include a certain number of generalizations relating different terms to each other, together with the laws of the propositional calculus. (Because of the 'inscrutability of terms'—cf. above pp. 140– 142—Quine does not think that predicate logic is susceptible of determinate translation. However I want at this point to consider those arguments for the indeterminacy of translation which are not based on the inscrutability of terms. So I shall ignore Quine's differential treatment of propositional and predicate logic in what follows.) Now it might seem that by requiring the preservation of stimulus analyticities Quine has included inferential connections between sentences after all. But there remain puzzles. For one thing, as the tortoise showed Achilles, we are not going to get any inferences from sentences alone, whatever those sentences say. We need at least some rules of inference as well. Still, let us grant Quine, as seems reasonable, that translation is supposed to preserve also those inferential procedures to which the laws of logic, and other stimulus-analytic generalizations, give expression. How does his position then fare?

In part this depends on how strictly the notion of stimulus-analyticity is to be taken. If it is taken very strictly, as requiring complete inviolability in the face of any empirical pressures whatsoever, then it is doubtful whether there are any stimulus-analyticities at all, at least not in scientific languages. No scientific generalizations have privileged immunity from empirically moti-vated revisions; even 'logically guaranteed' truths can succumb in various ways, as we saw earlier. So on a strict interpretation the stimulus-analyticity requirement does not get the inferential connec-tions between sentences back in after all. On the other hand we could interpret stimulus-analyticity less strictly, as only requiring relative tenacity in the case of empirical pressure, not absolute tenacity. This would seem more sensible, given (*contra* the *Word and Object* Quine) that our dispositions to use words non-inferentially in direct response to stimulations are themselves malleable. However, if we do so interpret stimulus-analyticity less strictly, then it is difficult to see how we can avoid counting all accepted lawlike generalizations as laying down rules of inference for other sentences containing their constituent predicates, and so as something translation must pre-serve. For where are we to draw the line? If it is merely relative tenacity in the face of counter-evidence that is required, then the fact that some generalizations have rather better credentials than others will not stop them all qualifying.

In any case, the indeterminacy thesis itself seems incoherent how-ever we take stimulus-analyticity. The idea, remember, was that we could have two (or more) translation manuals, both 'compatible with the totality of speech dispositions' yet taking a given sentence of the foreign language into two different English sentences which 'stand in no sort of equivalence however loose'. But how could this be? What is to make the alternative translations stand in 'no sort of equivalence however loose', unless it is that their use conforms to *different* 'speech dispositions'? But then how can both translations preserve the same 'totality of speech disposition'? Whatever we deem to be the relevant set of 'speech dispositions', surely this ought to go both for the evaluation of equivalence and for the constraining of translation. We can, as we have just seen, construe the 'totality of speech dispositions' more or less narrowly. If we required that translation preserve only the narrower class of 'speech dispositions', and used the wider class to evaluate 'equivalence', then of course we could have 'unequivalent' translations. But if it is indeed right to use the wider class for equivalence, then why should translation not have to preserve just that wider class? Quine seems to want to have his cake and eat it. (See

Boorse [1975] and Dummett [1974a] for this argument, and for comments on related points.)

We can see why Quine gets into this position. He appreciates, as perhaps some of his critics do not, that once we start relaxing our notion of what counts as a relevant speech disposition it is going to be very difficult to avoid letting in so much as to preclude any translation between two languages whose speakers hold different theories. So he inclines towards a tight notion of speech disposition. But then he quite rightly realizes that if that is all that has to be preserved then we are likely to end up with alternative translations which by any intuitive pre-theoretical standards are incompatible.

The holist theory of meaning grasps the nettle and admits that any theoretical differences will preclude translation. Of course we will often want to have some idea of what, in our terms, is meant by the words of an alien language set within a theoretical structure we do not share. And then a rendering which fails to preserve certain of the components of that theoretical structure will have to serve. But since all accepted generalizations contribute to meanings, this 'translation' should be recognized to fall short of the ideal. (In such a case it may well turn out that there are a number of mutually incompatible 'translations' of roughly the same degree of accuracy. But this would not be an example of 'divergent manuals' all 'compatible with the totality of speech dispositions'. The possibility of their divergence arises only because they are all perforce incompatible with that totality.)

5 HOLISM AND TRANSLATION

It may seem of little consequence whether we count the different possible ways of rendering the language of one theory into that of another as equally good translations or equally bad *mis*translations. However there are good reasons for preferring the latter description.

Suppose we see a translation as an aid to the understanding of an alien people's linguistic behaviour. A translation gives us an immediate way of identifying the contents of the various alien assertions, the beliefs that component speakers recognize those assertions to express. In doing this it shows us how to go about explaining why the aliens say what they do. (The central case will be where a speaker uses a sentence literally to express a belief he has been led to by appropriate evidence. But the explanations of assertions involving unliteralness, insincerity, or beliefs arrived at inappropriately will also require a grasp of what those sentences mean. Cf. p. 183 above.)

Now Quine's undemanding requirements on translation amount to the claim that the translation-relevant content of an assertion is exhausted by what is given by its direct association with sensory stimulation. In effect this denies that we can use a translation as a route to the explanation of any utterances other than those that issue directly from such sensory stimulations. But surely this is wrong, even if we prescind, in Quinean spirit, from questions of insincerity, unliteralness, inappropriately formed beliefs, etc. Suppose an alien community has the same theoretical structure as we have, framed in different words. Then the appropriate pairing of their words with ours will enable us to see directly not only how to explain their direct linguistic responses to sensory stimulation, but also how to explain those assertions which are arrived at by inferential reasoning within their theoretical structure. By identifying the beliefs behind their words in terms of theory-backed inferential procedures appropriate to the adoption of those beliefs, as well as any observational procedures which could warrant those beliefs, we will be in a position to explain far more of what the aliens say. Which is why a translation should be required to communicate the fuller 'holist' contents of foreign words rather than merely the emasculated 'Quinean' ones; and correlatively why alternative translations which alike fail to communicate these fuller contents perfectly should be deemed equally *bad* rather than equally good.

One possible source of confusion requires immediate comment. To say that a translation manual is an aid to the understanding of alien linguistic behaviour is not to say that such a manual is the only route to such an understanding. For there is always the possibility of interpreting an alien language without translating it: of grasping the contents of the alien expressions without being able to identify those contents with contents of expressions in the pre-existing home language. (Cf. above p. 74.) The holist requirement that translation should preserve all the inferential and observational procedures recognized by the aliens means of course that perfect translations will almost inevitably be impossible for languages at any cultural or historical distance from ours. But this does not mean that the linguistic behaviour of those who use(d) such languages must of necessity remain opaque to us, that we are precluded from ever understanding that behaviour. It merely means that the requisite grasp of the alien assertions' contents cannot be read straight off from a translation manual, but has to be reached by an independent route.

Still, even given this point, there remain difficulties with the holist requirement on translation. We are considering translation manuals as possible aids to the understanding of alien linguistic behaviour. The argument for the holist requirement on translation was that a translation manual satisfying it would enable us to explain any alien assertion that issued from conformity to their theory-backed inferential procedures, and not just those that emerged as direct observational responses. But what about the aliens' prior decisions on the generalizations which make up their theoretical structure and back their inferential procedures? What light can a holist translation manual cast on the acceptance or denial of such general assertions? On the holist account all their accepted generalizations play a part in fixing the meanings of the alien terms. So there seems no possibility of understanding alien decisions on generalizations as *resulting* from the meanings that our translation specifies for their terms. How then are such decisions to be explained?

The difficulty facing the holist approach here is the analogue of that discussed at the end of Chapter 4 and in the previous sections of this chapter. There we wondered how we could justify changes of generalizations if all such changes were meaning changes. The question of how to explain another community's changes of generalizations is in effect simply a transposition of the problem from the normative to the descriptive mode. By pursuing the problem in this transposed form we will be able to clarify a number of related but as yet unresolved issues.

It might seem to some that the holist approach to translation is guilty of excessive charity. The 'principle of charity' is often advocated as a means, in practice if not in principle, of avoiding the indeterminacy of translation: given a number of alternative translation manuals, all equally faithful to the 'objective reality' of linguistic use, choose the one which makes the aliens right in as much as possible of what they accept. So, according to the principle of charity, the right translation is the one which makes the aliens come out as correct as possible. But according to the holist theory of translation, no translation is right unless it makes the aliens come out as correct in every case where they are led to conclusions by their accepted observational and inferential procedures. By incorporating all those procedures into the meanings of the alien words the holist approach implies that all such conclusions must be warranted. And, correlatively, it implies that any translation which renders such an alien conclusion into an English sentence which is not warranted by the relevant evidence is in that respect defective. But surely this much charity on the part of a translator cannot be absolutely obligatory?

How damning is this accusation of excessive charity? We can best approach this question by evaluating the merits of the 'principle of charity' itself. On examination the principle of charity turns out to be at best a very rough approximation to the correct criteria for an optimal translation. And once we are clear on this point we will see that 'excessive charity' is in fact a somewhat ill-framed charge to bring against the holist approach to translation.

Translation is not just a matter of being nice to the natives. Whatever we hold to be the meanings that translations should preserve, surely the real test of a translation is that it indicates the right explanations of the natives saying what they do, not that it makes them seem correct in everything they say. As the discussion in Sections 4 and 5 of Chapter 3 should have made quite clear, there is nothing in itself wrong with a translation implying that certain of the natives' assertoric utterances are incorrect: our translational methodology needs to leave room for cases of insincerity, unliteralness, beliefs arrived at on inadequate grounds, etc. And, as we saw, this room can be left without making the selection of a translation manual arbitrary: for the imputations of incorrectness arising from a given translation are answerable to the availability of explanations of why the natives do in those cases say incorrect things. We can clarify the issue as follows. A translation manual indicates how to go about explaining the aliens' assertoric utterances. The central case is where the kind of evidence which the translation indicates as appropriate gives rise to the assertion in question. But this standard mode of explanation then carries with it the possibility of alternative cases, where the assertion is made in the absence of the accredited grounds. Such 'mistakes' are also quite explainable, in terms of insincerity, irrationality, etc. But provided they really are mistakes, provided they really are assertions which by the established alien standards are in actuality not warranted, then of course our translation should make them out as such, rather than attempt to turn them charitably into truths. So it is by no means a requirement of translation that it make the aliens come out right in everything (or in as much as possible of) what they say. It should do this only to the extent that they are so right. Of course the standards against which such rightness should be measured are those implicit in *their* language—for what in the end gives content to their sentences is what is laid down as adequate and appropriate grounds for asserting them. But to hold, as this implies, that translations should preserve such accredited grounds for assertions is not to hold that it should preserve the correctness of all assertoric utterances. For, to repeat, people do not always say what they should.

So indiscriminate translational charity is quite misguided. However, there is no real reason to think that the holist approach to translation is committed to the kind of translational charity just criticized. According to the holist the total set of generalizations accepted by a given community specifies what observational and inferential procedures can validate other sentences (in particular, singular sentences) involving the terms in those generalizations. And so he requires that translations of those terms should preserve that structure of accredited procedures. But it is no part of his theory that the alien speakers will never utter the relevant singular sentences except as a result of conformity to those procedures. The holist, like any translator, is quite free to hold that a certain assertion is the erroneous result of insincerity, or irrationality, etc., provided only that he can then defend his identification of the accredited procedures against which that assertion comes out erroneous, by finding independent evidence to back up his imputation.

So the trouble with the holist approach is not that it is overly charitable as such. It is as much able as any other approach to translation to allow that alien speakers can make individual utterances which belie the meanings which translation has to preserve. The real difficulty arises from its specific insistence that all accepted lawlike generalizations fix such meanings, and the corresponding requirement that a translation must map all the generalizations the aliens accept into ones we accept. For what this does imply is that an alien community can never be mistaken in accepting a generalization—or, rather better, that the question of a community's being so mistaken simply does not arise. In general, the decision to accept a sentence is warranted if it results from acquaintance with the accredited grounds indicated by that sentence's meaning, while it will be mistaken if it results from something else. But if the terms in generalizations do not have fixed meanings independently of decisions as to whether to accept those generalizations or not, then there will be no stable 'accredited grounds' by reference to which we might raise the question of whether such decisions are mistaken or not. And it is precisely because the holist approach so precludes any grip on the notion of a 'mistaken' decision to accept a generalization that it seems to leave us in the dark about how such decisions are to be explained. Since the holist model of meaning does not see decisions on generalizations as answerable to antecedently fixed meanings, it seems that a community's acceptance of a generalization simply has to be taken as an unexplainable datum: the lack of any fixed meaning precludes any identification either of the accredited evidence, acquaintance with which might have led to the generalization's adoption, or, alternatively, of what 'rationale' there might have been for accepting it even in the absence of such evidence.

There is thus a sharp contrast between the way the holist treats generalizations and the way he treats singular assertions. Decisions on singular assertions, unlike those on generalizations, do not disrupt the stable meanings to which those decisions are answerable. And there is thus no difficulty about distinguishing 'correct' from 'mistaken' singular assertions, nor about explaining these two kinds of singular assertions in the respectively appropriate ways.

There is of course nothing exceptionable about the holist attitude to singular sentences. Speakers do indeed recognize that their singular assertions are straightforwardly answerable to justification in terms of the relevant observational and inferential procedures. If such an assertion is accepted mistakenly, and then it later emerges that the required evidence is not after all available, then there is no question that we should cease to accept the sentence. In the terminology of Richard Grandy's [1973], we clearly have, in addition to the 'first-order' linguistic disposition manifested by our initially accepting the sentence, a *'second-order'* disposition to revise that acceptance when further evidence comes to light. (This is of course why the holist halts his holism when he gets to singular sentences. Even for the holist the acceptance of a singular sentence in itself adds nothing to the established principles governing a language—at best such an acceptance will only be a manifestation of the existing procedures for using the terms involved.)

But what does seem questionable is the holist's attitude to the acceptance of generalizations. Surely, it will be said, there must be some room for a linguistic community to be mistaken in accepting a generalization. Do we not ourselves repeatedly decide that we were wrong in our previous acceptance of certain generalizations? To return to an earlier example, surely we have now discovered that Newtonian physicists were wrong when they accepted the generalization 'light rays take different times to traverse congruent bodies with different motions'?

More specifically, is it not obvious that if we accept that singular assertions can be mistaken then we will also have to accept that generalizations can be mistaken? Suppose a generalization is inferred inductively from the mistaken acceptance of a series of singular sentences. Would this not be a clear case of a generalization being accepted wrongly? If it were discovered that the singular inductive grounds were mistaken, would it not be uncontroversial that the generalization was then to be rejected along with the singular sentences which gave rise to it?

However there is room for the holist to defend himself here. He could admit there is a sense in which such generalizations ought not to have been accepted in the first place. But he could nevertheless argue that once they are accepted as lawlike, they are as much part of the meanings of the terms they contain as any other generalizations—with the consequence that after their acceptance it ceases to make sense to think of them as 'mistakes'.

This holist response is admittedly rather *ad hoc*. But I shall not pursue this point any further here. For one thing, it seems unlikely that the kind of mistake in question ever does actually occur. Even if singular statements do on occasion get accepted mistakenly, it does not follow that this will happen uniformly enough to provide an adequate inductive basis for the acceptance of generalizations. And indeed, when we do consider examples where accepted singular sentences have yielded such an inductive basis, they do in general seem to have been warranted, up to their limits of accuracy, by the then accepted procedures for using them. (Thus the observed instances which fitted, say, the classical assumption 'mass is velocity invariant' were justified, up to the acknowledged limits of accuracy, by the classical procedures for using the terms 'mass' and 'velocity'.) In any case, the suggested danger of starting from a basis of mistaken singular sentences does not really get to the heart of the difficulty the holist has with generalizations. For even if we ignore this danger there still remains the inevitable inductive risk involved in accepting any generalization. With singular sentences about spatio-temporal particulars we can plausibly conceive of legitimate acceptance as requiring conclusive justification by experience. (Consider that from the holist perspective the legitimacy of singular sentences is best thought of relative to *currently accepted* procedures for using the relevant terms, in abstraction from the potential need to revise those procedures in the future. So the kind of conclusive justification in question should not be conceived of as showing 'truth', for, as we have seen, the indiscriminate inclusion of all currently accepted procedures makes any simple notion of truth inapplicable.) So we can take the mistaken acceptance of a singular sentence to require that the accepter lacks the experience that would conclusively justify it. But with generalizations we seem always to be in this position: notoriously, no amount of experiences can guarantee a universal generalization with potential future instances. However good the original grounds for acceptance, there always remains room for the future to discredit a previously accepted generalization. Does the holist not then have to admit that a generalization can turn out to be mistaken—namely, when despite having good inductive grounds, it is later discredited by a counter-example? Is this not exactly what happened, for example, with the classical principle instanced earlier: did not the Michelson-Morley experiment simply show that, despite previous indications, classical physicists were wrong to hold that 'light transit times for congruent bodies with different motions are different'?

But here too the holist can stick to his guns, and argue that this interpretation of the Michelson-Morley and similar 'falsifying' experiments is not uncontroversial. As an alternative he can suggest the interpretation according to which the classical mistake exposed by the Michelson-Morley experiment was not the one about 'light transit times' but instead the assumption that 'equal findings with rigid rods indicate congruence'. That is, he could take the classical terms involved to have had the meanings that would, so to speak, have imposed the 'Lorentz interpretation' on the Michelson-Morley experiment. For if the classical terms involved had really had those meanings then the right way for us, as relativity theorists, to describe the situation would indeed be to say that what the Michelson-Morley experiment showed was that equal rigid rod findings did not imply classical 'congruence'. This analysis would, it is true, demand an unfamiliar translation of the classical term 'congruent' into the language of special relativity, instead of the normal homophonic one (namely, that 'a and b are (classically) congruent' means that a and b have lengths in that ratio required for classical theory to predict that the light transit times would be equal). But given that the acoustic properties of the words involved are irrelevant in this context, this is scarcely an argument against the holist's proposed alternative analysis.

The holist's point is not of course that his alternative analysis is an improvement on the standard one. His aim is merely to call in question the unequivocal acceptance of the standard story. Let us rehearse the situation. Classical physicists accepted the following two sentences: 'the light transit times for congruent bodies with different motions are different' (sentence A) and 'equal findings with rigid rods indicate congruence' (sentence B). We have been presented with the choice of either translating standardly and so saying they were wrong to accept A, or translating non-standardly and saying their mistake was to accept B. The holist's point is that neither of these options is satisfactory. For to plump definitely for the first option is to imply that in the language of classical physics B had a privileged status by comparison with A—that the structure of the language predetermined that if and when evidence arose making it impossible to hold both A and B it would be A that went and B that remained inviolate. While the second option implies conversely that classical physicists were semantically predisposed to reject B and regard A as unthreatened by the envisaged evidence. But, as I have repeatedly argued, scientists do not discriminate amongst the components of their theoretical structure in the kind of way required for either of the suggested options to be right. There is nothing in classical physics to require that one of A and B was semantically inviolate (analytic) while the other was empirically underminable (synthetic), rather than that both were simply components of the same semantic kind in a total theoretical structure. So to opt for either translation is of necessity to misrepresent the language of classical physics in one way or the other. And, correspondingly, it would be wrong to pick out either A or B as a generalization that classical physicists were unequivocally 'mistaken' in accepting.

I do not of course want to deny that the Michelson-Morley experiment presented an anomaly for classical physics, that it demanded that the total theory be revised somewhere. Nor do I want to deny that in the end it was quite right for the sentence A to be rejected and for B to be retained. This admission might at first sight seem tantamount to allowing the standard translation of classical language after all. Does it not show that classical physicists *were* determinately predisposed to stand by B rather than A when a choice had to be made? And was this not just what was required for the standard translation?

However there are holes in this line of argument. For one thing, it is arguable that the move to relativity theory, and theoretical revisions in general, depend to a large extent on the contingencies of creative inspiration, and are by no means automatic products of the triggering by new evidence of determinate predispositions implicit in previous theoretical commitments. But even if we let that pass, and allow, implausibly, that new theories are always predetermined by previous commitments and current evidence, the predispositions involved would still not be such as to allow translation between the languages of different theories. For what will constrain a theory to be revised one way rather than another is not some supposed preferential semantic division amongst its constituent generalizations, but simply its overall structure in relation to the new evidence. In our example, it was not that classical physics was somehow originally committed to B more than to A, but simply that the best way to accommodate its total 'shape' to the anomaly presented by the Michelson-Morley experiment was by shifting A rather than B. If linguistic predispositions to revise generalizations in the face of new evidence were fixed by a sub-set of theoretical commitments then translation between the languages of different theories would indeed on occasion be possible, namely in those cases where the theories in question share those sub-sets of commitments. But as long as such predispositions have to be seen as fixed by overall theoretical structures, if by anything, then inter-theoretic translations will be impossible, and will only serve to mislead about the real content of the translated terms.

(If we *have* to translate between the languages of different theories, then there are good reasons for holding that a translation is the less misleading the more it preserves central truths rather than others. For the premium that is placed on the retention of central truths means that they are more important in structuring concepts, more significant in fixing what responses will be made to new evidence. But, as before, we should not forget that centrality is essentially a matter of degree, and so that even translations that preserve centrality will be only approximations to the ideal.)

In connection with the above points, it is worth spelling out that upholding, say, the Einsteinian revision of classical physics over the Lorentzian one, is not to accept the standard homophonic *translation* of the *language* of classical physics, nor even is it necessarily to rule out the 'Lorentzian translation': all that a non-standard interpretation of classical language need commit us relativity theorists to is a recognition that the move to relativity theory (quite rightly) changed the meanings of the relevant physical terms.

The difficulty which precipitated the discussion of the last few pages was that the holist approach to translation was apparently unable to yield explanations of an alien community's changes in accepted generalizations. Since the holist approach denied that generalizations have stable meanings across such changes, it could not account for such changes as simply resulting from new evidence in conformity to such meanings (nor even as resulting from illegitimate motivations which transgressed those meanings). But it should now have become clear that the holist does after all have a perfectly satisfactory way of explaining changes in generalizations, without supposing them to have stable meanings. For he can simply see such changes as due to the observance of general methodological principles. When a community is faced by an anomaly, they will be moved to look for some revision of their overall theory. And, assuming they do adhere to sound methodological principles, they will be further moved to look for that revision which promises most progress fastest. At the end of Chapter 4 I showed how methodological principles might justify one revision of a theory over others. In an entirely analogous way, a community's adherence to those methodological principles might explain why they in fact opted for that revision over the others. In particular, I argued at some length in Chapter 4 that it was perfectly consistent with all theory changes being meaning changes that different revisions of theory should be of different relative merit. Similarly here our ability to explain theory changes by reference to the observance of methodological principles does not in any way require that such changes be informed by stable meanings attaching to the generalizations involved.

Recall now the comments made on Zande talk about 'witches' ('ira mangu') in Section 4 of Chapter 3. I claimed there that in the standard case where a Zande judges that 'so and so is a witch' his judgment will be quite correct relative to the meaning of his term 'ira mangu', relative to his concept of a witch. The difference between Zande judgments and ours was to be attributed to their working with different concepts from ours, and not, as on the more standard view, to their applying the same concepts in illegitimate ways. We have now in effect filled out the argument for these claims. Since no particular bit of the Zande structure of general assumptions involving the concept of a witch can be picked out as the part which is mistaken, we have no option but to allow that all parts of that structure play a role in giving content to their concept of a witch. Consequently we should allow that singular judgments arrived at by conformity to the observational and inferential procedures backed by that theoretical structure arc faithful to the meaning of the term 'ira mangu'. (But note that this by no means implies that we should unreservedly endorse Zande talk of witches. We have good methodological reason to adopt quite different theoretical structures covering the phenomena in question, and so we will quite rightly refuse in the first place to engage seriously in the Zande discourse.)

There is a further point raised by traditional thought systems such as the Zande witchcraft theory. In the last paragraph but one I suggested that the holist can quite happily account for a community's changing its system of accepted generalizations by reference to its adherence to general methodologies principles, rather than in terms of its observing the dictates of stable meanings. But this assumes of course that the community in question does at some level recognize the requirements of scientific methodology. Yet clearly this can scarcely be taken for granted for all linguistic communities. In particular, though it is by no means absurd to claim that traditional thought systems are in general developed in accord with proper methodological precepts, there remain good reasons for doubting that they in fact are so developed. Such anthropological evidence as there is seems rather to suggest that traditional thinkers are inclined under pressure to let their theories degenerate *ad infinitum*, responding to any apparent anomalies by *ad hoc* devices which have the eventual effect of cutting off their theories entirely from any serious contact with the empirical phenomena. Nor is it plausible to suppose that methodological shortcomings are restricted solely to traditional societies. Even in 'scientifically mature' modern societies distorting pressures of one kind and another can prevent theories being developed in the methodologically optimal way. (See my [1978], Chs. 6 and 7, for a more detailed discussion of methodological deviance in traditional and modern societies.)

However the possibility of methodological shortcomings scarcely presents a serious problem for the holist. For he can simply explain methodologically errant theoretical revisions by reference to whatever factors are responsible for that errancy. Changes (or rather the relative lack of change) in traditional theories can be explained in terms of the relevant society's institutional lack of interest in methodological propriety, while distortions of modern theoretical development can be accounted for by reference to the specific pressures behind such socially uncharacteristic methodological contrariness. The holist sees good and bad theoretical revisions as alike changing meanings—but he has no difficulty in discriminating between them and explaining each in the appropriate manner.

To round off this set of comments on translation and related issues it is worth making clear that my reasons for denying the possibility of inter-theoretic translation are nothing to do with the underdetermination of theory. One of Quine's arguments for the indeterminacy of translation is premissed on a thesis of the underdetermination of theory by all possible empirical evidence. Quine holds that even given all the empirical data that could in principle be gathered, from the infinite past to the infinite future, there could still be alternative theories of reality between which it would be impossible to decide. This is the underdetermination of theory. Quine then argues that even if we knew all there was to know about the contents of some alien community's terms, all there was to know about how they were disposed to revise their current theory in response to possible future evidence, we would still not have fixed a translation for their terms. For there would still be a number of alternative possible translations, each presenting the aliens as (prospectively) committed to a different one of the set of alternative theories compatible with all possible future evidence. (See in particular Quine [1970].)

Quine's line of reasoning here seems valid. So, since the indeterminacy of translation makes no sense, it follows that the underdetermination of theory makes no sense either. I suspect that the underdetermination thesis gains what plausibility it has from a confused attitude to the double language model's observation-theory distinction. If it is assumed that there is a sharp and significant distinction between observational claims and theoretical ones then it indeed seems reasonable to allow that the totality of observational truths would still allow different theoretical assumptions. But then if we do allow such a sharp distinction it becomes difficult, as we saw in Chapter 1, to take theoretical talk all that seriously; and it becomes unclear in what sense the alternatives really are different theories and not just terminological variants of the same substantial claims. On the other hand, if we reject the observation-theory distinction, then it ceases to be plausible in the first place that theories are underdetermined by all possible evidence. Once we deny that there is a distinct category of 'theoretical' statements which get evidentially assessed only at second remove, then there remains no obvious reason for thinking that any epistemological slack will remain when all possible evidence is in. So theoretical differences seem compatible with the totality of possible evidence only on a view of 'theory' which makes those differences insubstantial. Once again Quine seems to want to have his cake and eat it.

Given that I have rejected the observation-theory distinction, I see no reason to deny that the totality of possible evidence will select a uniquely correct theory. But from my point of view this is no argument for the possibility of inter-theoretic translation. Remember that in practice we will always be short of that totality of evidence which will determine the perfect theory (cf. above p. 132). So what we must consider are the relationships between different imperfect theories, for they are all we will ever have. Now (granting for the sake of argument that everybody always develops their theories with perfect methodological propriety) it is true that people with different theories will tend to move towards the same 'final' theory as more evidence comes in. Given the standard conception according to which theoretical changes are constrained by stable meanings, it would indeed follow from this that they must attach the same meanings to certain of their terms. But on our account theory changes are not constrained by any commitments to meanings, but simply by the requirements of a meaning-independent methodology. So the fact that theoretical rivals in the end get led in the same direction does nothing to establish the existence of common semantic commitments behind their theoretical differences.

BIBLIOGRAPHY

Achinstein, P. [1968], *Concepts of Science*, Johns Hopkins Press, Baltimore.

Ayer, A. [1956], *The Problem of Knowledge*, Macmillan, London.

Boorse, C. [1975], 'The Origins of the Indeterminacy Thesis', *Journal of Philosophy*, LXXII, 369–87.

Braithwaite, R. [1953], *Scientific Explanation*, Cambridge University Press.

Brandom, R. [1976], 'Truth and Assertibility', *Journal of Philosophy*, LXXIII, 137–49.

Carnap, R. [1928], *Der logische Aufbau der Welt*, Berlin.

—— [1936], 'Testability and Meaning', *Philosophy of Science*, 3 (1936), 419–71, 4 (1937), 1–40.

Craig, E. [1976], 'Sensory Experience and the Foundations of Knowledge', *Synthese*, 33, 1–24.

Craig, W. [1953], 'On Axiomatizability Within a System', *Journal of Symbolic Logic*, XVIII, 30–2.

—— [1956], 'Replacement of Auxiliary Expressions', *Philosophical Review*, LXV, 38–55.

Davidson, D. [1967], 'Truth and Meaning', *Synthese*, 17, 304–23.

Dummett, M. [1958], 'Truth', *Proceedings of the Aristotelian Society*, LIX, 141–62.

—— [1973], *Frege: Philosophy of Language*, Duckworth, London.

—— [1974a], 'On the Significance of Quine's Indeterminacy Thesis', *Synthese*, 27, 351–97.

—— [1974b], *The Justification of Deduction*, O.U.P. for the British Academy, Oxford.

—— [1976], 'What is a Theory of Meaning II?', in *Truth and Meaning*, ed. G. Evans and J. McDowell, Clarendon Press, Oxford.

Evans, G. and McDowell, J. [1976], 'Introduction', in *Truth and Meaning*, ed. G. Evans and J. McDowell, Clarendon Press, Oxford.

Feyerabend, P. [1958], 'An Attempt at a Realistic Interpretation of Experience', *Proceedings of the Aristotelian Society*, LVIII, 143–70.

—— [1962], 'Explanation, Reduction, and Empiricism', in *Minnesota Studies in the Philosophy of Science, II*, ed. H. Feigl, M. Scriven and G. Maxwell, University of Minnesota Press, Minneapolis.

—— [1963], 'How to be a Good Empiricist', in *Philosophy of Science, The Delaware Seminar, II*, ed. B. Baumrin, Interscience Publishers, New York.

—— [1965a], 'Problems of Empiricism', in *Beyond the Edge of Certainty*, ed. R. Colodny, Prentice-Hall, Englewood Cliffs, N.J.

—— [1965b], 'Replies to Criticism', in *Boston Studies in the Philosophy of Science, II*, ed. R. Cohen and M. Wartofsky, Humanities Press, New York.

—— [1965c], 'On the "Meaning" of Scientific Terms', *Journal of Philosophy*, LXII, 266–74.

—— [1970a], 'Against Method', in *Minnesota Studies in the Philosophy of Science, IV*, ed. M. Radner and S. Winokur, University of Minnesota Press, Minneapolis.

—— [1970b], 'Consolations for the Specialist', in *Criticism and the Growth of Knowledge*, ed. I. Lakatos and A. Musgrave, Cambridge University Press.

—— [1975], *Against Method*, New Left Books, London.

Field, H. [1973], 'Theory Change and the Indeterminacy of Reference', *Journal of Philosophy*, LXX, 462–81.

Fine, A. [1967], 'Consistency, Derivability, and Scientific Change', *Journal of Philosophy*, LXIV, 231–40.

Frege, G. [1892], 'On Sense and Reference', in *Translations from the Philosophical Writings of Gottlob Frege*, ed. P. Geach and M. Black, 1970, Blackwell, Oxford.

Gillies, D. [1972], 'Operationalism', *Synthese*, 25, 1–24.

Grandy, R. [1973], 'Reference, Meaning and Belief', *Journal of Philosophy*, LXX, 439–52.

Gregory, R. [1970], *The Intelligent Eye*, Weidenfeld and Nicolson, London.

—— [1974], 'Perceptions as Hypotheses', in *Philosophy of Psychology*, ed. S. C. Brown, Macmillan, London.

Grice, H. [1957], 'Meaning', *Philosophical Review*, LXVI, 377–88.

—— [1968], 'Utterer's Meaning, Sentence Meaning and Word Meaning', *Foundations of Language*, 4, 1–18.

Grünbaum, A. [1960], 'The Duhemian Argument', *Philosophy of Science*, XXVII, 75–87.

—— [1971], 'Can We Ascertain the Falsity of a Scientific Hypothesis?', in *Observation and Theory in Science*, ed. M. Mandelbaum, Johns Hopkins Press, Baltimore.

Hanson, N. [1958], *Patterns of Discovery*, Cambridge University Press.

Harding, S., ed. [1976], *Can Theories be Refuted?*, Reidel, Dordrecht.

Harman, G. [1973], *Thought*, Princeton University Press.

—— [1975], 'Language, Thought and Communication', in *Minnesota Studies in the Philosophy of Science, VII*, ed. K. Gunderson, University of Minnesota Press, Minneapolis.

Hempel, C. [1965], *Aspects of Scientific Explanation*, Free Press, New York.

—— [1966], *Philosophy of Natural Science*, Prentice-Hall, Englewood Cliffs, N.J.

Hesse, M. [1974], *The Structure of Scientific Inference*, Macmillan, London.

Kripke, S. [1972], 'Naming and Necessity', in *Semantics of Natural Language*, ed. D. Davidson and G. Harman, Reidel, Dordrecht.

Kuhn, T. [1962], *The Structure of Scientific Revolutions*, University of Chicago Press.

—— [1970a], 'Postscript', in *The Structure of Scientific Revolutions, 2nd edn.*, T. Kuhn, University of Chicago Press.

—— [1970b], 'Reflections on My Critics', in *Criticism and the Growth of Knowledge*, ed. I. Lakatos and A. Musgrave, Cambridge University Press.

Lakatos, I. [1970], 'Falsification and the Methodology of Scientific Research Programmes', in *Criticism and the Growth of Knowledge*, ed. I. Lakatos and A. Musgrave, Cambridge University Press.

—— [1971], 'History of Science and its Rational Reconstructions', in *PSA 1970. Boston Studies in the Philosophy of Science, VIII*, ed. R. Buck and R. Cohen, Reidel, Dordrecht.

Lakatos, I. and Zahar, E. [1976], 'Why Did Copernicus's Programme Supersede Ptolemy's?', in *The Copernican Achievement*, ed. R. Westman, University of California Press, Los Angeles.

Lewis, D. [1969], *Convention*, Harvard University Press, Cambridge, Mass.

Mackie, J. [1962], 'Counterfactuals and Causal Laws', in *Analytical Philosophy, First Series*, ed. R. Butler, Blackwell, Oxford.

Martin, M. [1971], 'Referential Variance and Scientific Objectivity', *British Journal for the Philosophy of Science*, 22, 17–27.

Nagel, E. [1961], *The Structure of Science*, Routledge and Kegan Paul, London.

Papineau, D. [1978], *For Science in the Social Sciences*, Macmillan, London.

Peacocke, C. [1976], 'Truth Definitions and Actual Languages', in *Truth and Meaning*, ed. G. Evans and J. McDowell, Clarendon Press, Oxford.

Popper, K. [1959], *The Logic of Scientific Discovery*, Hutchinson, London.

Putnam, H. [1965], 'What Theories are Not', in *Logic, Methodology and the Philosophy of Science*, ed. E. Nagel, P. Suppes and A. Tarski, Stanford University Press.

—— [1969], 'Is Logic Empirical?', in *Boston Studies in the Philosophy of Science, V*, ed. R. Cohen and M. Wartofsky, Reidel, Dordrecht.

—— [1973a], 'Meaning and Reference', *Journal of Philosophy*, LXX, 699–711.

—— [1973b], 'Explanation and Reference', in *Conceptual Change*, ed. G. Pearce and P. Maynard, Reidel, Dordrecht.

—— [1975], 'What is Realism?', *Proceedings of the Aristotelian Society*, LXXVI, 175–94.

Quine, W. v. O. [1951], 'Two Dogmas of Empiricism', *Philosophical Review*, LX, 20–43.

—— [1960], *Word and Object*, M.I.T. Press, Cambridge, Mass.

—— [1969], 'Epistemology Naturalized', in *Ontological Relativity and Other Essays*, W. v. O. Quine, Columbia University Press, New York.

—— [1970], 'On the Reasons for Indeterminacy of Translation', *Journal of Philosophy*, LXVII, 178–83.

Quinton, A. [1955], 'The Problem of Perception', *Mind*, 67, 28–51.

Scheffler, I. [1967], *Science and Subjectivity*, Bobbs-Merrill, New York.

Schlick, M. [1936], 'Meaning and Verification', *Philosophical Review*, XLV, 339–69.

Sellars, W. [1963], *Science, Perception and Reality*, Routledge and Kegan Paul, London.

Shapere, D. [1966], 'Meaning and Scientific Change', in *Mind and Cosmos*, ed. R. Colodny, University of Pittsburgh Press.

Tuomela, R. [1973], *Theoretical Concepts*, Springer-Verlag, Berlin.

Wright, C. [1976], 'Truth Conditions and Criteria', *Aristotelian Society Supplementary Volume L*, 217–45.

Zahar, E. [1973], 'Why Did Einstein's Research Programme Supersede Lorentz's?', *The British Journal for the Philosophy of Science*, 24, 95–123 and 223–62.

INDEX